AF540691

Cruise Geography

Cruise Geography

Medwin Gale

RANDOM PUBLICATIONS
NEW DELHI (INDIA)

Cruise Geography

ISBN 978-93-5111-933-3

Published in 2016 in India by

RANDOM PUBLICATIONS

4376-A/4B, Gali Murari Lal, Ansari Road
New Delhi-110 002
Phone : +9111-43580356, 011-23289044, 011-43142548
e-mail: sales@randompublications.com,
info@randompublications.com, randomexports@gmail.com

Reprinted 2024

Type Setting by : Friends Media, Delhi-110089
Digitally Printed at : Replika Press Pvt. Ltd.

Preface

The cruise ship industry is currently undergoing a period of rapid expansion. New cruise capacity threatens to produce overtonnaging, with future berths exceeding demand. Despite this development, cruise operators are confident that a growing North American market will be able to satisfy the equilibrium condition sought by vessel operators. This research explores where the new tonnage might be deployed and its eventual impact on the cruise industry.

The modern cruise industry is also one of the most outstanding examples of globalization, with an increasing number of ports of call and destinations around the globe, a multinational clientele and onboard personnel from every continent, and a level of detachment from communities and nations never seen before in history, with important economic, legal, environmental and social implications.

A crisis-resistant industry with a diversified offer of airlift options and modernized port structures that have open up cruising as a vacation alternative available for an increasing, more affluent customer base, offering a exciting, eventful, relaxing and definitely enjoyable experience for millions of passengers from across the world each year.

This book considers the influence and effect of geography on the cruise industry. To start with, it is impossible to consider cruising without reflecting on the conditions that arise from the prevailing climate. Passenger comfort and safety are directly affected if a cruise ship sails in a particular part of an ocean or sea at a particular time of the year. This also holds true for destinations visited and shore activities that may be offered.

– Author

Contents

1

Geography of Cruise Ships

Cruising on Cruise ships is a means of travel with some substantial benefits, but a few drawbacks too. The majority of people love them, but some people hate them. Cruising makes it easy to visit several places in a single trip without the need to repack your belongings and sit in a car/train/bus/plane to travel to each one; your hotel room comes along with you, and even provides the transportation.

Additionally, seeing several islands or cities in a region can help you decide if and where you'd want to visit later for a longer time. Typical itineraries also limit the time you can spend in each place, usually just a short day of activities or sightseeing. They may also include one or more days at sea: paradise if you enjoy a relaxing day by the pool, but frustrating if you prefer more active and open exploration. non-etheless, the benefits far outweigh the drawbacks for enough people to support a growing industry.

Today you can visit every continent on earth, including Antarctica, by cruise ship. The most exotic itineraries, such as the Galapagos Islands, are best visited by small expedition vessels. While these cruises are expensive, you'll be traveling with expert guides and excursions are usually included. As you may note already, this article focuses on ocean cruising and ships.

A parallel article (to be developed) would focus on river boat and barge canal cruising. River boats and barges offer more in-depth, close-up looks at many countries in their interiors. For travel on smaller vessels, see Cruising on small craft.

Carnival Corporation is the giant in the ocean cruise industry. It owns Carnival Cruise Lines, Princess Cruises, Holland America, Cunard Line, Costa Cruises and Seabourn Cruises. The other major cruise lines are Royal Caribbean International, which owns Celebrity Cruises and Azamara Cruises, Oceania Cruises and P and O, which caters to the British market, and Norwegian Cruise Lines, which caters primarily to passengers on the United States' east coast with year round sailings from New York City and Miami. Cruise ship holidays have become are part of global geography and are also big business! It is difficult to give an exact number to the size of the global cruise ship fleet because there are many of different sizes and locations. There are small ones on rivers like the Nile, Rhine and Danube and huge ones on the oceans, but an estimate would be more than 300 ocean vessels and thousands of smaller river boats and touring yachts.

Fig. Cruise Ship Holidays have Become a Part of Global Geography and are also Big Business!

Old style cruise ships, or 'liners', were narrower-shaped and more boat-like for big ocean crossings, for example the QE2 which carried 1600 passengers, seen here on her last voyage in 2006.

Fig. The QE2.

New style ships like this one in Bergen Harbour, Norway, can carry over 2000 people - a floating town - and there are even bigger ones like the wrecked Costa Concordia with 3700 guests! Each ship has hundreds of cabins, verandas and windows; some have water parks and even ice rinks! They tend to concentrate their travel in tours of particularly attractive coastlines.

Fig. The Costa Atlantica.

So is this tourism all good news for the places they visit and the planet?

GEOGRAPHY

The port city of Fremantle lies on a series of limestone hills at the mouth of the Swan River, approximately 25 minutes from Perth, the capital of Western Australia. Many of the city's most impressive heritage buildings are built from the pale-coloured local limestone. Central Fremantle is bounded to the north and east by the lower reaches of the Swan River and to the west by the Indian Ocean with stunning stretches of beach to the north of the city and to the south. Fremantle is a major port for Western Australia and Fremantle Harbour is the entrance point for the many container ships, cruise ships, ferries and leisure craft that berth at the port in the Inner Harbour. Smaller leisure craft and ferries travel the 20 km's up the Swan River to Perth. Cycle paths along the southern side of the Swan River extend to Perth city and visitors can easily see rocky limestone formations along the route.

A series of islands lie off the coast of Fremantle including the famous holiday island of Rottnest, Garden Island and Carnac Island. Rottnest Island is the only island that has a regular ferry service transporting visitors to and from Fremantle. The enclosed and sheltered Fishing Boat Harbour and Success Harbour protect the city's working and recreational fleets and are within walking distance of the city centre. The Fishing Boat Harbour precinct also features boardwalks, café's restaurants and a brewery.

Fremantle has over 70 parks and reserves, from native bush land, beach parks and sporting reserves. Some of Fremantle's favourite parks include:

- Esplanade Park–a beautiful inner city park surrounded by tall Norfolk pine trees that sits on reclaimed land within the city centre.
- Monument Hill Reserve–a stunning WWII memorial and a spectacular lookout spot with views over the city, ocean and river.
- Fremantle Park–a popular public park, used by local groups for sport and recreation.

- Samson Park–the largest and most significant bush land reserve on the outskirts of the city is Fremantle's only Bush Forever park.
- Booyeembara Park–this stunning 16 hectare park developed by the City of Fremantle in conjunction with the local community, is an open space of regional significance and sustainability.

CRUISE SHIP

A cruise ship or cruise liner is a passenger ship used for pleasure voyages, where the voyage itself and the ship's amenities are a part of the experience, as well as the different destinations along the way. Transportation is not the prime purpose, as cruise ships operate mostly on routes that return passengers to their originating port, so the ports of call are usually in a specified region of a continent. There are even "cruises to nowhere" or "nowhere voyages" where the ship makes 2–3 night round trips without any ports of call.

Fig. Majesty of the Seas.

By contrast, dedicated transport oriented ocean liners do "line voyages" and typically transport passengers from one point to another, rather than on round trips. Traditionally, an ocean liner for the transoceanic trade will be built to a higher standard than a typical cruise ship, including high freeboard and stronger plating to withstand rough seas and adverse conditions encountered in the open ocean, such as the North Atlantic. Ocean liners also usually have larger capacities for fuel, victuals, and other stores for consumption on long voyages, compared to dedicated cruise ships.

Although often luxurious, ocean liners had characteristics that made them unsuitable for cruising, such as high fuel consumption, deep draught that prevented their entering shallow ports, enclosed weatherproof decks that were not appropriate for tropical weather, and cabins designed to maximize passenger numbers rather than comfort (such as a high proportion of windowless suites). The gradual evolution of passenger ship design from ocean liners to cruise ships has seen passenger cabins shifted from inside the hull to the superstructure with private verandas. The modern cruise ships, while sacrificing qualities of seaworthiness, have added amenities to cater to water tourists, and recent vessels have been described as "balcony-laden floating condominiums".

Fig. Cruise Ships in Tallinn Passenger Port of Tallinn,Estonia - a Popular Tourist Destination

The distinction between ocean liners and cruise ships has blurred, particularly with respect to deployment. Differences in construction remain. Larger cruise ships have also engaged in longer trips such as transoceanic voyages which may not return to the same port for months (longer round trips). Some former ocean liners operate as cruise ships, such as Marco Polo. This number is diminishing. The only dedicated transatlantic ocean liner in operation as a liner (as of December 2013) is Queen Mary 2 of the Cunard fleet. She also has the amenities of contemporary cruise ships and sees significant service on cruises.

Cruising has become a major part of the tourism industry, accounting for U.S.$29.4 billion with over 19 million passengers carried worldwide in 2011. The industry's rapid growth has seen nine or more newly built ships catering to a North American clientele added every year since 2001, as well as others servicing European clientele. Smaller markets, such as the Asia-Pacific region, are generally serviced by older ships. These are displaced by new ships in the high growth areas. The world's largest cruise ship is currently Royal Caribbean International's Allure of the Seas beating her sister ship (Oasis of the Seas) by about 2 inches.

HISTORY

Origins

The birth of leisure cruising began with the formation of the Peninsular and Oriental Steam Navigation Company in 1822. The company started out as a shipping line with routes between England and the Iberian Peninsula, adopting the name Peninsular Steam Navigation Company. It won its first contract to deliver mail in 1837. In 1840, it began mail delivery to Alexandria, Egypt, via Gibraltar and Malta. The company was incorporated by Royal Charter the same year, becoming the Peninsular and Oriental Steam Navigation Company.

Fig. SS Strathaird, a P and O Cruise Ship of the Early 20th Century. The Company Began Offering Luxury Cruise Services in 1844.

P and O first introduced passenger cruising services in 1844, advertising sea tours to destinations such as Gibraltar, Malta and Athens, sailing from Southampton. The forerunner of modern cruise holidays, these voyages were the first of their kind, and P and O Cruises has been recognised as the world's oldest cruise line.

The company later introduced round trips to destinations such as Alexandria andConstantinople. It underwent a period of rapid expansion in the latter half of the 19th century, commissioning larger and more luxurious ships to serve the steadily expanding market. Notable ships of the era include the SS Ravenna built in 1880, which became the first ship to be built with a total steel superstructure, and the SS Valetta built in 1889, which was the first ship to use electric lights.

Some sources mention Francesco I, flying the flag of the Kingdom of the Two Sicilies (Italy), as the first cruise ship. She was built in 1831 and sailed from Naples in early June 1833, preceded by an advertising campaign. The cruise ship was boarded by nobles, authorities, and royal princes from all over Europe. In just over three months, the ship sailed to Taormina, Catania, Syracuse, Malta, Corfu, Patras,Delphi, Zante, Athens, Smyrna, Constantinople and delighting passengers with excursions and guided tours, dancing, card tables on the deck and parties on board. However, it was restricted to the aristocracy of Europe and was not a commercial endeavour.

The cruise of the German ship Augusta Victoria in the Mediterranean and the Near East from 22 January to 22 March 1891, with 241 passengers including Albert Ballin and wife, popularized the cruise to a wider market. Christian Wilhelm Allers published an illustrated account of it as Backschisch (Baksheesh).

Fig. Prinzessin Victoria Luise was the First Ship Purpose-Built as a Cruise Ship.

The first vessel built exclusively for luxury cruising, was Prinzessin Victoria Luise of Germany, designed by Albert Ballin, general manager of Hamburg-America Line. The ship was completed in 1900. The practice of luxury cruising made steady inroads on the more established market for transatlantic crossings. In the competition for passengers, ocean liners added luxuries — Titanic being the most famous example — such as fine dining, luxury services, and staterooms with finer appointments. In the late 19th century, Albert Ballin, director of the Hamburg-America Line, was the first to send his transatlantic ships out on long southern cruises during the worst of the winter season of the North Atlantic. Other companies followed suit. Some of them built specialized ships designed for easy transformation between summer crossings and winter cruising. In 1896, there were three luxury liners for transportation, for the Europe to North America trip. These were European-owned. In 1906, the number had increased to seven. The British Inman Line owned City of Paris, the Cunard Line had Compania and Lucania. The White Star Line owned Majestic and Teutonic. La Lorraine and La Savoiewere owned by the French Compagnie Générale Transatlantique.

Modern Luxury Cruising

Fig. Queen Elizabeth 2 was Reinvented as a Luxury Ocean Liner Following the Advent of the Jet Airliner.

With the advent of large passenger jet aircraft in the 1960s, intercontinental travelers switched from ships to planes sending the ocean liner trade into a terminal decline. Certain characteristics of older ocean liners made them unsuitable for cruising duties, such as high fuel consumption, deep draught preventing them from entering shallow ports, and cabins (often windowless) designed to maximize passenger numbers rather than comfort. Ocean liner services aimed at passengers ceased in 1986, with the notable exception of transatlantic crossings operated by the British shipping company Cunard Line, catering to a niche market of those who appreciated the several days at sea. In an attempt to shift the focus of the market from passenger travel to cruising with entertainment value, Cunard Line pioneered the luxury cruise transatlantic service on boardthe Queen Elizabeth 2 ocean liner. International celebrities were hired to perform cabaret acts onboard and the crossing was advertised as a vacation in itself.

Queen Elizabeth 2 also inaugurated "one-class cruising" where all passengers received the same quality berthing and facilities. This revitalized the market as the appeal of luxury cruising began to catch on, on both sides of the Atlantic. The 1970s television series Love Boat, helped to popularize the concept as a romantic opportunity for couples.Another ship to make this transition was SS Norway, originally the ocean liner SS France and later converted to cruising duties as the Caribbean's first "super-ship".

Fig. Freedom of the Seas, Formerly the Largest cruise Ship in the World.

Contemporary cruise ships built in the late 1980s and beyond, such as Sovereign-class which broke the size record held for decades byNorway, show characteristics of size and strength once reserved for ocean liners—some have undertaken regular scheduled transatlantic crossings.

The Sovereign-class ships were the first "megaships" to be built, they also were the first series of cruise ships to include a multi-story atrium with glass elevators. They also had a single deck devoted entirely to cabins with private balconies instead of oceanview cabins. Other cruise lines soon launched ships with similar attributes, such as the Fantasy-class and Crown Princess. As theveranda suites were particularly lucrative for cruise lines, something which was lacking in older ocean liners, recent cruise ships have been designed to maximize such amenities and have been described as "balcony-laden floating condominiums".

Fig. Oasis of the Seas in Nassau,Bahamas.

Until 1975-1980, cruises offered shuffleboard, deck chairs, "drinks with umbrellas and little else for a few hundred passengers." After 1980, they offered increasing amenities. As of 2010, city-sized ships have dozens of amenities. There have been nine or more new cruise ships added every year since 2001, all at 100,000 GT or greater. The only comparable ocean liner to be completed in recent years has been Cunard Line's Queen Mary 2 in 2004. Following the retirement of her running mate Queen Elizabeth 2 in November 2008, Queen Mary 2 is the only liner operating on transatlantic routes, though she also sees significant service on cruise routes. Queen Mary 2 was for a time the largest passenger ship before being surpassed by Royal Caribbean International's Freedom-class vessels in 2006. The Freedom-class ships were in turn overtaken by RCI's Oasis-class vessels which entered service in 2009 and 2010.

OPERATORS AND CRUISE LINES

Fig. Celebrity Solstice in Port Melbourne,Australia.

Operators of cruise ships are known as cruise lines. Cruise lines have a dual character; they are partly in the transportation business, and partly in the leisure entertainment business, a duality that carries down into the ships themselves, which have both a crew headed by the ship's captain, and a hospitality staff headed by the equivalent of a hotel manager. Among cruise lines, some are direct descendants of the traditional passenger shipping lines (such as Cunard), while others were founded from the 1960s specifically for cruising.

Historically, the cruise ship business has been volatile. The ships are large capital investments with high operating costs. A persistent decrease in bookings can put a company in financial jeopardy. Cruise lines have sold, renovated, or renamed their ships to keep up with travel trends. The cruise lines operate their ships virtually 24 hours a day, seven days a week, 52 weeks a year. A ship which is out of service for routine maintenance means the loss of tens of millions of dollars. If the maintenance is unscheduled, it can result, potentially, in thousands of dissatisfied customers.

A wave of failures and consolidation in the 1990s has led to many lines existing only as "brands" within larger corporations, much as a single automobile company produces several makes of cars. Brands exist partly because of repeat customer loyalty, and also to offer different levels of quality and service. For instance, Carnival Corporation owns both Carnival Cruise Lines, whose former image were vessels that had a reputation as "party ships" for younger travelers, but have become large, modern, yet still profitable, and Holland America Line, whose ships cultivate an image of classic elegance. In 2004, Carnival Corporation had merged Cunard's headquarters with that of Princess Cruises inSanta Clarita, California so that administrative, financial and technology services could be combined, ending Cunard's history where it had operated as a standalone company (subsidiary) regardless of parent ownership. However, Cunard did regain some independence in 2009 when its headquarters were moved to Carnival House in Southampton.

Some cruise lines have specialties; for example, Saga Cruises only allows passengers over 50 years old aboard their ships, and Star Clippers and formerly Windjammer Barefoot Cruises and Windstar Cruises only operate tall ships. Regent Seven Seas Cruises operates medium-sized vessels—smaller than the "megaships" of Carnival and Royal Caribbean—designed such that 90 per cent of their suites are balconies. Several specialty lines offer "expedition cruising" or only operate small ships, visiting certain destinations such as the Arctic and Antarctica, or the Galápagos Islands.

Currently the five largest cruise line operators in the world are Carnival Corporation and plc, Royal Caribbean Cruises Ltd., Star Cruises (which owns 50 per cent of Norwegian Cruise Line; NCL in its own right is the third largest line), MSC Cruises, and Louis Cruise Lines. Louis Cruises has largely grown its fleet through purchasing older second- or third-hand ships, while the other four operators have largely constructed their own vessels and combined own the majority of the "megaships".

ORGANIZATION

Fig. Disney Magic.

Fig. AIDAdiva in Izmir.

Cruise ships are organized much like floating hotels, with a complete hospitality staff in addition to the usual ship's crew. It is not uncommon for the most luxurious ships to have more crew and staff than passengers.

Dining

Dining on almost all cruise ships is included in the cruise price. Traditionally, the ships' restaurants organize two dinner services per day, early dining and late dining, and passengers are allocated a set dining time for the entire cruise; a recent trend is to allow diners to dine whenever they want. Having two dinner times allows the ship to have enough time and space to accommodate all of their guests.

Having two different dinner services can cause some conflicts with some of the ship's events (such as shows and performances) for the late diners, but this problem is usually fixed by having a shorter version of the event take place before late dinner. Cunard Line ships maintain the class tradition of ocean liners and have separate dining rooms for different types of suites, while Celebrity Cruises and Princess Cruises have a standard dining room and "upgrade" specialty restaurants that require pre-booking and cover charges. Many cruises schedule one or more "formal dining" nights. Guests dress "formally", however that is defined for the ship, often suits and ties or even tuxedos for men, and formal dresses for women. The menu is more upscale than usual. Besides the dining room, modern cruise ships often contain one or more casual buffet-style eateries, which may be open 24 hours and with menus that vary throughout the day to provide meals ranging from breakfast to late-night snacks. Ships also feature numerous bars and nightclubs for passenger entertainment; the majority of cruise lines do not include alcoholic beverages in their fares and passengers are expected to pay for drinks as they consume them. Most cruise lines also prohibit passengers from bringing aboard and consuming their own beverages, including alcohol, while aboard. Alcohol purchased duty-free is sealed and returned to passengers when they debark.

There is often a central galley responsible for serving all major restaurants aboard the ship, though specialty restaurants may have their own separate galleys. As with any vessel, adequate provisioning is crucial, especially on a cruise ship serving several thousand meals at each seating. For example, passengers and crew on theRoyal Caribbean International ship Mariner of the Seas consume 20,000 pounds (9,000 kg) of beef, 28,000 eggs, 30,000 L of ice cream, and 18,000 slices of pizza in a week. Normally, a cruise ship stocks up at its home port. They also have special arrangements with designated suppliers at ports of call if required.

Other on-Board Facilities

Most modern cruise ships feature the following facilities:

- Casino — Only open when the ship is at sea to avoid conflict with local laws
- Spa
- Fitness center
- Shops — Only open when ship is at sea to avoid merchandising licensing and local taxes
- Library
- Theatre with Broadway-style shows
- Cinema
- Indoor and/or outdoor swimming pool with water slides
- Hot tub
- Buffet restaurant
- Lounges
- Gym
- Clubs
- Basketball courts
- Pool tables
- Ping pong tables

Some ships have bowling alleys, ice skating rinks, rock climbing walls, sky-diving simulator, miniature golf courses, video arcades, ziplines, surfing simulators, basketball courts, tennis courts, chain restaurants and/or ropes obstacle courses.

Fig. H_2O Zone aboardFreedom of the Seas.

Fig. Promenade on the Allure of the Seas.

Fig. Golf course on Brilliance of the Seas.

Fig. Molecular Bar aboard theCelebrity Equinox before Christmas.

Crew

Crew is usually hired on three to eleven month contracts which may then be renewed as mutually agreed, which is based upon service ratings from passengers as well as the cyclical nature of the cruise line operator. Most staff work 77 hour work weeks for 10 months continuously followed by 2 months of vacation. There are no paid vacations or pensions for service, non-management crew, depending on the level of the position and the type of the contract. Non-service and management crew members get paid vacation, medical, retirement options, and can participate in the company's group insurance plan. The direct salary is low for North American standards, though restaurant staff have considerable earning potential from passenger tips. Crew members do not have any expenses while on board as food and accommodation, and transportation for most employees, are included. This makes a cruise ship career financially attractive enough to compensate for lack of employment benefits.

Living arrangements vary by cruise line, but mostly by shipboard position. In general two employees share a cabin with a shower, commode and a desk with a television set, while senior officers are assigned single cabins. There is a set of facilities for the crew separate from that of passengers, such as mess rooms and bars, recreation rooms, prayer rooms/mosques, and fitness center, with some larger ships even having a crew deck with a swimming pool and hot tubs.

For the largest cruise operators, most "hotel staff" are hired from under-industrialized countries in Asia, Eastern Europe, the Caribbean, and Central America. While several cruise lines are headquartered in the United States, the ships are registered in countries such as Bahamas, Panama, and Liberia, a practice known as "flags of convenience" to take advantage of less-stringent labour regulations. Collective action is difficult due to staff being from diverse cultural and ethnic backgrounds. The exception to this are the cruise routes around Hawaii, where operators are required to register their ships in the United States and the crew is unionized, so these cruises are typically much more expensive than Caribbean and Mediterranean.

Business Model

The "luxury cruise lines" such as Regent Seven Seas Cruises and Crystal Cruises provide "the most all-inclusive" cruises. Base fare on Regent Seven Seas ships includes most alcoholic beverages onboard ship and most shore excursions in ports of call, as well as all gratuities that would normally be paid to hotel staff on the ship. Fare also includes one night's hotel stay before boarding and airfare to and from the cruise's origin and destination ports. Most cruise lines since the 2000s have priced the cruising experience, to some extent, a la carte, as passengers spending aboard generates significantly more from ticket sales. The passenger's ticket includes the stateroom accommodation, room service, unlimited meals in the main dining room and buffet, access to shows, and use of pool and gym facilities. However, there are extra charges for alcohol and soft drinks, official cruise photos, Internet and wi-fi access, and specialty restaurants; it has been reported that the casino and photos have high profit margins. Cruise lines earn significantly from selling onshore excursions (keeping 50 percent or more of what passengers spend for these tours) offered by local contractors.In addition, cruise ships earn significant commissions for sales from onshore stores that are promoted on board as "preferred" (as much as 40 percent of gross sales).

Facilitating this practice are modern cruise terminals with establishments of duty-free shops inside a perimeter accessible only by passengers and not locals. Ports of call have often oriented their own businesses and facilities towards meeting the needs of visiting cruise ships. In one case, Icy Strait Pointin Alaska, the entire destination was created explicitly and solely for cruise ship visitors. Travel to and from the port of departure are the passengers' responsibility, although purchasing a transfer pass from the cruise line for the trip between the airport and cruise terminal will guarantee that the ship will not leave until the passenger is aboard. Similarly, if the passenger books a shore excursion with the cruise line and the tour runs late, the ship is obligated to remain until the passenger returns.

Ship Naming

Older cruise ships have had multiple owners. it is usual for the transfer of ownership to entail a refitting and a name change. Some ships have had a dozen or more identities. Many cruise lines have a common naming scheme they use for their ships. Some lines use their name as a prefix or suffix in the ship name (such as the prefixes of "Carnival", "AIDA", "Disney", or "Norwegian" and the suffix of "Princess"). Other lines use a unique word or phrase (such as the prefix of "Pacific" for P and O Cruises Australia or the suffixes of "of the Seas" for Royal Caribbean International or "-dam" for ships of the Holland America Line). The addition of these prefixes and suffixes allows multiple cruise lines to use the same popular ship names while maintaining a unique identifier for each ship.

Cruise Ships Utilization

Due to slower speed and reduced seaworthiness, as well as being largely introduced after several major wars, cruise ships have never been used as troop transport vessels. By contrast, ocean liners were often seen as the pride of their country and used to rival liners of other nations, and have been requisitioned during both World Wars and theFalklands War to transport soldiers and serve as hospital ships. Cruise ships and former liners often find employment in applications other than those for which they were built. A shortage of hotel accommodation for the 2004 Summer Olympics led to a plan to moor a number of cruise ships in Athens to provide tourist accommodation.

On 1 September 2005, the U.S. Federal Emergency Management Agency (FEMA) contracted three Carnival Cruise Lines vessels (Carnival Fantasy, the former Carnival Holiday, and the Carnival Sensation) to house Hurricane Katrina evacuees. In 2010, in response to the shutdown of UK airspace due to the eruption of Iceland's Eyjafjallajökull volcano, the newly completed Celebrity Eclipse was used to rescue 2000 British tourists stranded in Spain as an act of goodwill by the owners. The ship departed from Southampton for Bilbao on 21 April, and returned on 23 April.

REGIONAL INDUSTRIES

Fig. Four Ships at the Cruise Ship Terminal in Nassau, The Bahamas.

Fig. Cruise Ships in Ushuaia, Argentina.

Most cruise ships sail the Caribbean or the Mediterranean. Others operate elsewhere in places like Alaska, the South Pacific, the Baltic Sea and New England. A cruise ship that is moving from one of these regions to another will commonly operate a repositioning cruisewhile doing so. Expedition cruise lines, which usually operate small ships, visit certain more specialized destinations such as the Arctic and Antarctica, or the Galápagos Islands.

The number of cruise tourists worldwide in 2005 was estimated at some 14 million. The main region for cruising was North America (70 per cent of cruises), where the Caribbean islands were the most popular destinations. The second most popular region was continental Europe (13 per cent), where the fastest growing segment is cruises in the Baltic Sea. The most visited Baltic ports are Copenhagen, St. Petersburg, Tallinn, Stockholm and Helsinki. The seaport of St. Petersburg, the main Baltic port of call, received 426,500 passengers during the 2009 cruise season. According to 2010 CEMAR statistics the Mediterranean cruise market is going through a fast and fundamental change; Italy has won prime position as a destination for European cruises, and destination for the whole of the Mediterranean basin. The most visited ports inMediterranean Sea are Barcelona (Spain), Civitavecchia (Italy), Palma (Spain) and Venice (Italy). 2013 saw the entrance of the first Chinese company into the cruise market. China's first luxury cruise ship, Henna, made her maiden voyage from Sanya Phoenix Island International Port in late January.

Caribbean Cruising Industry

Fig. Nearly 9,000 Passengers from Three Carnival Ships Visiting St. Thomas, US Virgin Islands; from Front to Back:Carnival Liberty, Carnival Triumph and the Carnival Glory.

The Caribbean cruising industry is one of the largest in the world, responsible for over $2 billion in direct revenue to the Caribbean islands in 2012. Over 45,000 people from the Caribbean are directly employed in the cruise industry. An estimated 17,457,600 cruise passengers visited the islands in the 2011-2012 cruise year (May 2011 to April 2012.) Cruise lines operating in the Caribbean include Royal Caribbean International, Princess Cruises, Carnival Cruise Line, Celebrity Cruises, Disney Cruise Line, Holland America,P and O, Cunard, Crystal Cruises, Pullmantur Cruises and Norwegian Cruise Line. There are also smaller cruise lines that cater to a more intimate feeling among their guests. The three largest cruise operators are Carnival Corporation, Royal Caribbean International, and Star Cruises/Norwegian Cruise Lines.

Some of the American cruise lines in the Caribbean depart from ports in the United States, "nearly one-third of the cruises sailed out ofMiami". Other cruise ships depart from Port Everglades (in Fort Lauderdale), Port Canaveral (approximately 45 miles (72 km) east ofOrlando), New York, Tampa, Galveston, New Orleans, Cape Liberty, Baltimore, Jacksonville, Charleston, Norfolk, Mobile, and San Juan, Puerto Rico. Some UK cruise lines base their ships out of Barbados for the Caribbean season, operating direct charter flights out of the UK and avoiding the sometimes lengthy delays at US immigration. The busiest ports of call in the Caribbean for cruising in the 2013 year are listed below

Rank	Destination	Passenger Arrivals (2013)[39]
1	Bahamas	4,709,236
2	Cozumel, Mexico	2,751,178
3	United States Virgin Islands	1,998,579
4	Sint Maarten	1,779,384
5	Cayman Islands	1,375,872
6	Jamaica	1,288,184
7	Puerto Rico	1,176,343
8	Turks and Caicos Islands	778,920
9	Aruba	688,568
10	Belize	677,350
11	Haiti	643,634
12	Saint Kitts and Nevis	629,000
13	Curacao	610,186
14	Saint Lucia	594,118
15	Barbados	570,263
16	Antigua and Barbuda	533,993
17	Dominican Republic	423,910
18	British Virgin Islands	367,362
19	Bermuda	320,090
20	Dominica	230,588
21	Grenada	197,311
22	Martinique	103,770
23	Bonaire	96,818
24	Saint Vincent and the Grenadines	82,974

SHIPYARDS

The construction market for cruise ships is dominated by three European companies and one Asian company:

- STX Europe of South Korea with one shipyard:
- STX France at Chantiers de l'Atlantique in Saint-Nazaire, France.
- Mitsubishi Heavy Industries of Japan.
- Meyer Werft of Germany with two shipyards:
- Meyer Werft of Germany.
- Meyer Turku at Perno shipyard in Turku, Finland
- Fincantieri of Italy.

A large number of cruise ships have been built by other shipyards, but no other individual yard has reached the large numbers of built ships achieved by the four above. A handful of old ocean liners also remain in service as cruise ships.

SAFETY

Security

As most of the passengers on a cruise are affluent and have considerable ransom potential, not to mention a considerable amount of cash and jewelery on board (casino and shops), there have been several high profile pirate attacks on cruise ships, such as on the Seabourn Spirit and MSC Melody.

As a result, cruise ships have implemented various security measures. While most merchant shipping firms have generally avoided arming crew or security guards for reasons of safety, liability and conformity with the laws of the countries where they dock, cruise ships have small arms (usually semi-automatic pistols) stored in a safe accessible only by the captain who distributes them to authorized personnel such as security or the master-at-arms. The ship's high-pressure fire hoses can be used to keep boarders at bay, and often the vessel itself can be maneuvered to ram pirate craft. A recent technology to deter pirates has been the LRAD or sonic cannon which was used in the successful defence of Seabourn Spirit.

Passengers entering the cruise ship are screened by metal detectors. Explosive detection machines used include X-ray machines and explosives trace-detection portal machines (a.k.a. "puffer machines"), to prevent weapons and contraband on board. Security has been considerably tightened since September 11, 2001, such that these measures are similar to airport security. In addition to security checkpoints, passengers are often given a ship-specific identification card, which must be shown in order to get on or off the ship. This prevents people boarding who are not entitled to do so, and also ensures the ship's crew are aware of who is on the ship. The Cruise Ship ID cards are also used as the passenger's room key. CCTV cameras are mounted frequently throughout the ship.

Stability

The design of cruise ships has changed dramatically during the past decades. One of the biggest changes has been moving the passenger cabins from inside the hull to the superstructure and adding individual balconies both due to customer demand and because, from a business standpoint, the cruise line can charge passengers much more than for inside staterooms. This has considerably increased the overall height of the ships, making them more susceptible to side wind and waves. As a result, there have been concerns about the stability of modern passenger ships especially in heavy weather. Because there is much more ship above the surface than beneath it, modern cruise ships may appear top-heavy to some. Despite the large superstructure, the center of mass of modern cruise ships is relatively low. This is due to large open spaces and the extensive use of aluminium, high-strength steel and other lightweight materials in the upper parts, and the fact that the heaviest components — engines, propellers, fuel tanks and such — are located in the lower parts of the ship. Thus, even though modern cruise ships may appear top-heavy, proper weight distribution ensures they are not. Furthermore, large cruise ships also tend to be very wide, which considerably increases their initial stability by increasing the metacentric height.

Although most passenger ships utilize stabilizers to reduce rolling in heavy weather, they are only used for crew and passenger comfort and do not contribute to the overall intact stability of the vessel. The ships must fulfill all stability requirements even with the stabilizer fins retracted.

Safety Record

From 2005 to the Costa Concordia disaster in 2012, out of more than 100,000,000 people worldwide who had taken cruises, there had been 16 fatalities.

INFECTIONS ON CRUISE SHIPS

Norovirus

Norovirus is the most common cause of gastroenteritis in developed countries and is so widespread that only the common cold is reported more frequently. Symptoms usually last between 1 and 3 days and generally resolve without treatment or long term consequences. The incubation period of the virus averages about 24 hours.

The estimated likelihood of contracting gastroenteritis from any cause on an average 7-day cruise is less than 1 per cent. In 2009, during which more than 13 million people took a cruise, there were nine reported norovirus outbreaks on cruise ships. Outbreak investigations by the U.S. Centers for Disease Control and Prevention have shown that transmission among cruise ship passengers is primarily person-to-person; potable water supplies have not been implicated. In 2009, ships undergoing unannounced inspections in U.S. ports received an average CDC Vessel Sanitation Programme score of approximately 97 out of a total possible 100 points. The minimum passing inspection score is 85. Collaboration with the CDC's Vessel Sanitation Programme and the development of Outbreak Prevention and Response Plans have helped to generally decrease the incidence of norovirus outbreaks on ships. The CDC states that the reasons why noroviruses are associated with cruise ships is because "(1) Health officials track illness on cruise ships. So outbreaks are found and reported more quickly on a cruise ship than on land; (2) Close living quarters may increase the amount of group contact; and (3) New passenger arrivals may bring the virus to other passengers and crew." As of 15 February 2013, data for the year 2012 shows that there were 16 reported incidents of norovirus illnesses including an outbreak on Cunard Line's 'Queen Mary 2.'

Legionnaires' Disease

Other pathogens which can colonise pools and spas including those on cruise ships include Legionella, the bacteriium which causes Legionnaires' disease. Legionella, and in particular the most virulent strain, Legionella pneumophila serogroup 1, can cause infections when inhaled as an aerosol or aspirated. Individuals who are immunocompromised and those with pre-existing chronic respiratory and cardiac disease are more susceptible. Legionnaires' has been infrequently associated with cruise ships. The Cruise industry Vessel Sanitation Programme has specific public health requirements to control and prevent Legionella.

Enterotoxigenic Escherichia Coli (ETEC)

Enterotoxigenic Escherichia coli is a form of E. Coli and the leading bacterial cause of diarrhea in the developing world, as well as the most common cause of diarrhea for travelers to those areas. Since 2008 there has been at least one reported incident each year of E. Coli on international cruise ships reported to the Vessel Sanitation Programme of the Centers for Disease Control. Causes of E. Coli infection include the consumption of contaminated food and the ingestion of water in swimming pools contaminated by human waste.

ENVIRONMENTAL IMPACT

Fig. Diesel Smoke from Cruise Ships Over Juneau, Alaska.

"Cruise ships generate a number of waste streams that can result in discharges to the marine environment, including sewage, graywater, hazardous wastes, oily bilge water, ballast water, and solid waste. They also emit air pollutants to the air and water. These wastes, if not properly treated and disposed of, can be a significant source of pathogens, nutrients, and toxic substances with the potential to threaten human health and damage aquatic life. Cruise ships represent a small — although highly visible — portion of the entire international shipping industry, and the waste streams described here are not unique to cruise ships. Particular types of wastes, such as sewage, graywater, and solid waste, may be of greater concern for cruise ships relative to other seagoing vessels, because of the large numbers of passengers and crew that cruise ships carry and the large volumes of wastes that they produce. Because cruise ships tend to concentrate their activities in specific coastal areas and visit the same ports repeatedly (especially Florida, California, New York, Galveston, Seattle, and the waters of Alaska), their cumulative impact on a local scale could be significant, as can impacts of individual large-volume releases (either accidental or intentional)." Some cruise lines, such as Cunard, are taking steps to reduce environmental impact by refraining from discharges (RMS Queen Mary 2 has a zero-discharge policy) and reducing their CO2 output every year.

THE GLOBAL HONEY POTS FOR CRUISE SHIPS

Geographers use the term 'honey pot' to describe popular tourist destinations that cater for large numbers of visitors. The term is usually applied to small places, like a national park visitor centre in the Brecon Beacons, but some international coastal areas have attracted millions of tourists. Can you identify on this world map the names of the popular rivers and coastal / island cruising attractions? Of course some cruises go all round the World or all around the Pacific Ocean - these can visit over 100 places and take 115 or more days-one third of a year on the cruise!)

How does this rapidly growing cruise ship industry affect on the places they visit? Well, there is some good news and some bad news.

The Cruise ship industry has many good features:

- It provides very successful holiday packages for an estimated 20 million people per year.
- It creates employment and income for on-board workers and the ports they use throughout the world.

But cruise ships are attracted to the most beautiful and easily damaged natural environments. There are many who concerned about a non-ecotourism industry visiting ecologically sensitive areas.

Here are a few of the concerns that have been raised:

- They have a history of accidental damage to coral reefs and coastlines through collision and anchoring. Wreck of the Costa Concordia is an area that was unpolluted and rich in wildlife. In November 2007, two cruise ships sank in Antarctica. Many tons of fuel, hydraulic fluids, lubricants, harmful chemicals from televisions, computer screens, etc. were released into the protected Antarctic waters.

- Analysts calculate that a cruise liner such as Queen Mary 2 emits double the CO_2 per passenger mile, compared to commercial passenger aircraft.
- The fuel oil burned by ship engines is a heavy air pollutant making air quality in ports particularly poor as these pictures taken in Bergen, Norway show. The smoke over the city is all from just one ship! Throughout the world, they release an estimated 1.2 million to 1.6 million metric tons of tiny airborne particles each year.

Fig. Bergen, Norway.

- Because of the holiday and luxury nature of the lifestyle, passengers cruise ships each produce on average 3.5 kilograms of rubbish daily - compared with the 0.8 kilograms each generated by local people on shore. (Source: Our Planet)
- Visits to wildlife areas are very frequent and are unregulated for example Within one month (January) a survey team from the Scott Polar Research Institute in Antarctica recorded 14 visits by six tour ships with over 2000 tourists landing on one penguin island.
- A large cruise ship on a one week voyage is estimated to generate 210,000 gallons of human sewage and 1 million gallons (40 more swimming pools) of gray water (water from sinks, baths, showers, laundry and galleys). Cruise ships also generate large volumes of oily bilge water, sewage sludge, rubbish and hazardous wastes. Management of this waste could, with cost, be environmentally neutral but the industry has a poor record of pollution incidents of 'grey water' releases in to the delicate environments they visit.
- Cruise ships take on water as ballast (balancing weight) in one place and release it somewhere else later on when the ship's weight needs adjusting. This moves all sorts of plants and animals into new places and creates problems including that of invading species taking over.
- Cruises do provide the areas they visit with income, but most of the money goes to the shipping and tour company and not to the holiday destinations.

Fig. The Costa Concordia.

The Cruise industry is showing some efforts to increase efficiency, decrease pollution and use technology to manage waste, but it only takes an incident like the wreck of the Costa Concordia to show that the environment is still at risk from 'accidents' even from the largest, environmentally managed and well-equipped modern ships.

Understand

Fig. The Upper Deck of a Typical Cruise Ship.

Fig. Irozaki Cruise.

The golden age of transoceanic passenger travel is long gone, and the only surviving ships from that era are all either converted to cruising, preserved as museums and/or hotels, or are laid-up, but that doesn't mean that traveling across the sea by ship is gone too. In truth, modern-day passenger ships, including Cunard Line's mammoth Queen Mary 2, are actually now much larger and more luxurious than they were decades ago. The Cunard "Queens" still occasionally make traditional fast Atlantic crossings, *e.g.*, between New York and England. Reassuringly, most cruise ships seldom segregate public areas or restaurants that passengers may use based on the "quality" of the cabin they've purchased, as was widely customary in the early 20 century.

The picture of cruise ship travel painted by the circa-1977 TV series "The Love Boat" isn't particularlymisleading (except about the inevitable bliss before debarkation and the all-American crew), but it is ratherincomplete. Due to economy of scale, most modern cruise ships carry 2,000 to 5,000 passengers. While the luxury segment of the cruise industry boasts small ships...even "boutique" vessels or "mega-yachts"...you'll probably board a small floating city. Voyages range from a few days to full circumnavigations of the globe lasting three months, while fares range from a few hundred dollars to $100,000+. Premium and Luxury cruise lines tend to have much smaller capacity ships carrying 100-1250 or so passengers. Most ships carry 2,000-3,500 passengers, while mega-ships can carry over 5,000...a virtual city on something that weighs many times as much as the Titanic.

Each size has its merits, e.g.,:

- Smaller ships can visit smaller harbors in highly desirable or unusual locales, *e.g.*, the Riviera, Galapagos, rugged shorelines. They may also be able to port inside a city at an older city port, saving up to 2 hours of travel from where larger ships must dock.
- Larger ships may offer a few more amenities...as discussed in "On-board" below...but must use well-sized harbors (or anchor/moor off-shore) with transport and touring infrastructures to handle so many people.
- Mega-ships offer huge public spaces and wide-ranging activities, but are limited to major and sometimes even custom built ports...again with even greater infrastructures.

Cruise lines offer widely varying itineraries. Examples range from...

- A few days at sea or to a nearby port-of-call. These may be offered as an introduction to cruising, or just as an opportunity to party.
- One or two weeks to visit ports and sights in a particular region, per "Cruise types" below.
- A month or more to see a region of the world, or three or more months to go around the world; a lifetime experience.

Each cruise is comprised of one or more cruise segments, *e.g.*, a 1-2 week "round-tripper" will be one segment, while visiting two or more regions may involve 2-3 segments...sometimes of an around-the-world cruise. That way, cruise lines can sell affordable "pieces" of long cruises that otherwise few can consider. Not infrequently, cruisers will buy two or three back-to-back/

sequential segments to build a longer cruise, *e.g.*, 7-10 days from Florida to the Western Caribbean, return, then 7-10 days for the Eastern, or two or more world regions when they are far away.

Cruise Types

Fig. Inside a Silja Cruise Ship.

Your experience will be substantially affected by the cruise type you choose. On a port-intensive itinerary, except for a few sea days...

- You might dine one evening, then enjoy entertainment, dancing, etc., go to sleep, and wake-up docked at your next port of call.
- Under a typical full-day port visit, you can often start ashore at 7-8 AM, and be expected back 30 minutes or so before ship departure at 5-6 PM or so.
- You can breakfast at your place of choice and be off-ship on your way to an organized tour, self-arranged tour or activities, or just a walk-about.
- An "all-day" tour may include lunch (see tour description); a half-day tour can have you back to the ship for lunch...then further touring, shopping or a walk-about for the afternoon if you choose.
- If you prefer such cruises...
- Look for itineraries in regions that offer many stops near the port where your cruise starts, *e.g.*, Mediterranean, Baltic, The Fjords, United Kingdom, Caribbean, Alaska, Mexican Riviera.
- Look for departure ports within the region, to minimize sea days,

e.g., Caribbean trips starting in San Juan or Barbados, or Amsterdam or Copenhagen for the Baltic or Fjords.
- In contrast, ships from Miami, Fort Lauderdale or Tampa can take 1-2 days to reach the Caribbean, and the same to return. But you have more cruise and ship choices there, and you can choose to drive to either port if you live close enough (see "Get in" below).

If you prefer sea days, you can look for:
- Re-positioning cruises (often crossing oceans)...usually taking two weeks or more...often involving one-way international flights. (See "Flying..." under "Get In" below.)
- Distant island or region visits from a mainland port, *e.g.*, Hawaii from Los Angeles.
- Segments of around-the-world cruises, usually traversing major regions over a 3-6 weeks. They too will involve one-way international flights.
- Around-the-world cruises, taking 90+ days...best considered once you know you're comfortable with many days at sea.
- As you find temptation, look at "Do" below and the ship's description (on-line or in brochure) to appreciate on-board activities.

There are also various cruises for special interests — for bridge players with a few masters on board, celebrity entertainers/bands, cultural/political science/ history lectures, GLBT lifestyles, etc. There have even been Linux cruises and other geek cruises.

Cruise Seasons

Fig. Eidfjord Cruise.

Many cruising regions have "high", "low" and "shoulder" seasons. These usually track with the most and least desirable times to visit the region, and times in-between, *e.g.*, Winter for the Fjords, late-Summer/Fall for the Caribbean (tropical storms) are deemed undesirable. Expect to pay premium prices during high-season, substantially less in "low", and perhaps you'll find some bargains in "shoulder", *e.g.*, for "re-positionings". Many ships transfer among distant regions that have opposite high-seasons...under repositioning cruises, *e.g.*, between the Mediterranean, Baltic or Alaska and the Caribbean, South America, Australia, New Zealand. The long distances require many sea days, often at attractive per day prices for those who like them.

Fig. Charlotte Amalie Harbor.

Fig. Vancouver Cruise.

Cruise Demographics

While the cruise industry once catered primarily to seniors, the age of passengers has diversified significantly. The average age of Royal Caribbean's passengers is 48, many other lines appeal to 20-40 year-old couples, "party" cruises attract young adults, and Disney and others focus on families with children and teens. Cruising has turned into an enormously popular family vacation due to well-designed children's programmes, even special cabin configurations.

Some itineraries and cruise-lines may attract more seniors, *e.g.*, trans-Atlantic and -Pacific re-positioning cruises, Holland-America (it very effectively supports but doesn't just market to seniors). Reasons include cost, cruise length, numbers of days at sea, and dates that conflict with school. If cruise demographics are important to you, read the cruise description carefully, research web sites on cruising (see "Other resources" below), and work through your travel/cruise agent to learn the probable demographics of any trip you're considering. You'll be glad you did so, or you risk embarking on a ship filled with rowdy young adults or seniors with limited activities.

If Handicapped or Pregnant

If you have a physical limitation, the cruise line can usually help...especially if notified in-advance. Without sacrificing personal privacy, use your agent (or directly if no agent is involved) to let them know about your needs and when they apply, *e.g.*,:

- Help with the significant walking distances to embark and dis-embark (processs discussed below), or to go on port visits.
- If you purchase a fly/cruise or cruise-extension package (also discussed below), which other help you'll need.

A cruise can place you some distance from proper pre-natal care and birthing facilities, especially advanced medical care. If you have any complication, or might be well-into your third-trimester during a cruise under consideration, consult your doctor. Then inform the cruise-line, through your agent if used. A note of fitness for travel from him/her can assuage the fears of the line and staff as you go through embarkation processing. Many cruise lines will in fact not permit you to cruise once you reach a certain point in your pregnancy (similar to airlines). Check this with your travel agent as you might be denied boarding upon arrival to the pier if you are too far along with your pregnancy.

Booking a Cruise

Once you've found a cruise that suits you, you'll want to know the full costs and terms...including port fees and taxes, often not reflected in brochure and initial web-site prices.

- On a web-site, as you approach the "buy" point, you'll be given a quoted price for all the parameters and options (*e.g.*, flying to/from the port(s)) you've chosen. Before you "buy", ensure you make a full copy of the quote/invoice and fully understand what's involved. It needs to also include all terms of the cruise contract you are agreeing to.
- Per "Cabins", "Get in" and other discussions below, wise choices of those parameters/options can be quite complex. Unless you are a seasoned cruiser, you should seriously consider using an agent rather than booking on-line yourself. He/she can help you considerably to understand all options available and their import, as well as basic costs, port fees and taxes that may not be apparent in on-line information. The contract also usually obligates you to pay for unexpected cruise line costs, *e.g.*, fuel surcharges if the world oil market has inflated by cruise time. Before you buy the cruise, he/she will also provide an invoice reflecting all costs, as of that date,

for your careful examination, and allowing you to question details or request changes.

Travel Insurance

If your trip starts to gain complexity or substantial cost as you plan it, or you are a first-time cruiser, you should consider trip insurance. Example reasons include if you intend to go on "adventure tours" (with risk of injury), have any medical condition that could flare up and require evacuation, will be a great distance from home, any possibility that any provider of trip services might go bankrupt, or if you've been forced to accept tight airline connections.

You usually must buy it after you have purchased your trip...shortly after is best. Its cost will basically be determined by your total trip cost, the age of travelers in your group, levels of coverage, and options for coverage requested for certain problems, *e.g.*, treatment for sickness or injury (on and off the ship), or medical evacuation. As you purchase it, do not understate the known, fixed costs of the entire trip; otherwise, your claim may be denied. Many policies will cover pre-existing medical conditions if purchased within a few days of booking your trip; they often will not if bought later.

Travel experts often recommend getting insurance from other than airlines or cruise lines. Their coverage often focuses only or primarily on their responsibilities, while a quality policy will cover your end-to-end trip, and many more risks. Some "lines" also offer cancellation protection, but their cost can be a major fraction of those for well-purchased, overall trip coverage. You may obtain better rates and/or coverage by buying it through or from an association you belong to, *e.g.*, AAA, AA.

For an in-depth discussion, see Travel insurance.

Key Ships Officers

The ship and your cruise depend on them. Just a few of them include:

- *The Captain:* He or she is called the ship's master for a reason, with total operational command of the vessel and when and where it goes. And yes, through recent changes to law, many can now officiate at weddings, as can accredited, "resident" or pre-certified clergy.
- *The Hotel Manager:* In charge of all staff that deliver on-board services, *e.g.*, Pursor/Customer Services, food and bar service, cabin staff, tour office, shops and nearly countless behind-the-scenes support staff.
- *The Cruise Director:* Responsible for all entertainment, special activities, key briefings and announcements, and any port or shopping advisers.
- The Maitre de for your dining room, the headwaiter, and (perhaps) most-especially your waiter for the area where your table is located if you have opted for traditional evening dining (see "Eat" below). They stand ready to make special arrangements for you on request

such as birthday or anniversary celebrations, table changes, and special or required dining needs, *e.g.*, food allergies, special diets.

Travel or Cruise Agent

As noted throughout this article, a knowledgeable agent can greatly help you obtain the best value for your cruise, and avoid many kinds of mistakes and avoidable travel missteps. He/she can/should learn your needs and preferences, and then know how to find what best fits your plans/preferences to discuss them with you before booking. Ask neighbors who travel or look near home for an agency. Be selective and you'll often find someone who can help at remarkably moderate cost... and for complex trips worth every penny.

Other Resources

Several web-sites provide objective information about various cruise lines, ships, cruising regions and ports, and how to choose, prepare for and go on a cruise. Many offer professional reviews, some offer passenger reviews. But because they often sell cruises through third parties, they cannot be listed here. To find them, use a good search engine, with "cruise" and "advice" or "review" as keywords among your search parameters.

Those sites and travel magazines discuss other valuable topics, *e.g.*, "wave season" (when to book, not when to go) versus other times, understanding what's included (and not) in prices shown, industry trends that may cause prices to go down. A good travel/cruise agent will have those and other insights. Knowing exactly when and how to best book a cruise receives nearly constant attention in travel articles, and approaches being an art.

Cabins

Fig. The Upper Deck of a Typical Cruise Ship.

Your accommodations can range widely...usually determined by cost. Most cruise lines promote their ships as luxurious, and cabin (aka stateroom, not room) furnishings can range from quite "nice" to "utterly elegant". The less expensive tend to be quite a bit smaller than ordinary hotel rooms...space you may only use for a few hours each day to sleep anyway. But every square inch is usable, *e.g.*, luggage fits under the bed to allow you to unpack many/all items and hang them in closets or store on shelves/in drawers for easy access.

Cabin grades/categories. On large ships, you'll find a number of cabin grades or categories within each cabin type. They involve location, size, quality of view, features, etc. Good travel/cruise agents have access to the codes for the nuances of features and know of shortfalls for many. Cabin costs will vary not only type but by those gradations/categories. For any cabin type, costs reflected in brochures and on web sites usually apply to the lowest grade.

Cabin types The basic types include:

- Inside cabins are the least expensive and located in the interior of the ship. Hence the name, they lack a window, but are filtered with the ship's air ventilation "piped-in".
- Ocean view... with windows (don't open, ship's air ventilation "piped-in")...slightly more expensive. The least expensive may have partially or substantially obstructed views.
- Balcony/veranda at even higher prices, with outside chairs, perhaps a table, to watch sunsets, have a room-service meal or treat, and watch passing ships and land.
- Mini-suites and full suites (the latter often multi-room) with private verandas. shower/bathtubs, sitting areas, perhaps hot tubs and other amenities...to command the highest prices.

Perhaps oddly, suites and the least expensive cabins tend to sell out first. Some cabins and all hallways have handrails for safety during occasional rough weather...not often needed. On ships built in the late 1990s or later, very few passengers will be:

- Bothered by pitching and rolling of the ship...all built with highly-effective stabilizers.
- Disturbed by the ship's engines or screws (propellers)...very quiet.
- Disturbed by frequent public announcements...easily heard in hallways and public areas. Except for key announcements, they are usually not piped into cabins, but can be heard on a designated TV channel.

Features:

- Virtually all cabins have twin beds, usually joined to create a generous queen, with side tables/drawers or shelves. (See "Sleep" below.) Suites may have king-sized beds.
- Cabins configured for families may also have a pull-down or wall-mounted bunk-bed, sleeper sofa or settee, or another twin/queen bed.
- All come with a small safe. While on-board, you should lock all valuables in it (*e.g.*, fine jewelry, passports, charge cards, cash), and leave them there unless needed, *e.g.*, for a port visit, shopping ashore or dressing for dinner.

- Expect to find a large wall-mounted mirror or two. Handy for checking your appearance, they also make the cabin seem bigger.
- Small private bathrooms with showers are the minimum, with better cabins offering more space, shower/baths or larger showers. Each type will offer at least minimal toiletries typical for a motel (if you need specific ones, bring them), small cabinets and shelves for all toiletries, at least one counter and lavatory, and atoilet that operates by power suction.
- All will have a phone with wake-up call capability (synchronized to the ship's time).
- Virtually all will have a TV, some even an attached DVD player.
- All will have at least a reach-in closet with a hanging rod, some hangars and a shelf often holding your life-vests. You'll also find storage drawers elsewhere. Suites may have walk-ins, with numerous shelves.
- Better rooms...uh, cabins...may have a settee, desk with chair or more.
- Better ships/cabins often offer a small refrigerator, holding chilled cans and bottles for sale. If you bring your own drinks, ask the cabin steward to empty it of items for sale.
- Power outlets and wattage are minimized...essential to avoid fire risk.
- The bathroom should have a low-wattage, usually 120V 60Hz outlet adequate for such as a shaver, and often a 220V outlet for the ship's hair-dryer.
- If not in the bathroom, the 220V outlet may be near the bed or desk, perhaps with the ship's hair-dryer stored nearby.
- Near the bed or desk, you may find one or two low-wattage, usually 120V 60Hz outlets adequate only for such as a laptop.
- Net result: Don't use your own iron or hairdryer or anything that heats food or liquid.
- Knowing these features and limits, experienced cruisers variously bring:
- An extension cord to use low-wattage 120 volt items at convenient locations, and 1-2 nite-lites.
- A small, plug-in surge arrester to prevent damage to personal electronics.
- For warm climes, a small fan. Cabin air conditioners effectively but slowly change cabin temperature, but airflow is often modest.
- Some form of air deodorizer...balcony fresh air (if you have one) quickly helps but can compromise cabin air temperature and humidity.
- For passengers with mobility challenges, some cabins have wide doors, "friendly thresholds into bathrooms and showers/tubs, and other helpful provisions.

Location can affect price somewhat because parts of a ship are more desirable to some passengers, e.g.,:

- To avoid the effects of ship's rolling or pitching, some may opt for a cabin on a lower deck or closer to amidship.
- To sunbathe on their balcony, many choose a deck well-below any over-hanging topdeck.
- Those who need quiet to sleep will choose locations away from lobbies and elevators, and with at least one deck between their cabin and any place where late-night revelers celebrate.
- Those with mobility challenges may prefer to be near elevators.

Cabin water is fully potable...usually obtained by reverse-osmosis...so efficient that some large ships visiting ports with water shortages may offload potable water. Older ships may use distillation supplemented by fresh water on-loads; distillation tends to make the water softer. All ships carefully treat the water to ensure its safety. Taste in cabins may be somewhat bland or have a hint of chemicals. Elsewhere, water often receives additional filtration to assure excellent taste for use in bars, dining rooms, kitchens, and buffet self-serve drink dispensers.

If a ship offers laundromats (usually consistent across a cruise line), each usually has more than one pair of washers/dryers and one or more irons and ironing boards usable only there. You'll also find detergent and softener dispensers.

Machines and dispensers usually require coins or tokens...obtained at the Pursor's desk, perhaps from a coin machine in the laundry room. As you get interested in any cruise, ship or cabin type, go to the cruise line's web site and others (*e.g.*, [1]) for more details. Again, a good travel/cruise agent can help you find the features you need or want.

GET IN

Fig. Golden Princess of the Princess Cruise Line
Docked at the Cruise Terminal at Basseterre, Saint Kitts.

Fig. Cruise Ships in Miami.

The best-known destinations for cruise ships are tropical ports in the Caribbean or the Mexican Riviera, theMediterranean and Northern Europe, but cruises can be found almost anywhere there's enough water to float a ship and cities or sites to visit. Cruise ships of various sizes visit the coasts of Alaska, Scandinavia,South-East Asia, East Asia, southern Europe, Australia and New Zealand, Oceania and New England; and various islands of the Pacific Ocean. Even the North Pole and Antarctica are now destinations, though the latter has emerging ecological questions. In addition, specially designed river boats and barges ply navigable rivers and lakes of Europe, China,Brazil, Egypt, North America and numerous other places. However, as noted above, this article focuses on ocean cruising and ships.

What to Pack

This can vary substantially according to the region you'll cruise, *e.g.*, clothing for cool/cold areas versus warm, conservative colours for Europe, items to cover arms and legs to enter many religious buildings worldwide. For more discussion, see Packing for a cruise, and especially the many good articles at travel, destination and cruise-oriented web-sites.

If you'll fly to/from a cruise port, see Fundamentals of flying for advice and suggestions. Many experienced cruisers find certain items necessities, *e.g.*,:

- Sunglasses with strong light filtration, to include full UV protection.
- Sanitizing wipes or solution. If you purchase any, look for active sanitizing ingredients beyond just alcohol. Most intended for hands have a glycerin base to prevent drying skin. Avoid using them on hard surfaces as they leave an undesirable residue. You'll note numerous uses below and in the separate article "Fundamentals of flying".
- Skin protection. Essential for all cruises, the sun not only strikes you from above, but is reflected off the water, *e.g.*, on boat excursions,

at beaches. (See full article on sunburn and protection at [2]) Pack and use protection, *e.g.*, adequate clothing, brimmed hats, sunblock with high UV A/B protection ratings. Ship's stores have some such items but charge dearly for them. You'll also find them ashore, but usually at a premium over home stores.

- Binoculars. Most views from ship are at a considerable distance.
- Duct tape. Useful to make temporary repairs to luggage and other items.
- Power Strip. Most smaller cruise ship cabins have one power outlet in the open cabin, and one in the lavatory. A strip becomes useful if you bring a laptop or cellphone (to recharge batteries), electric shaver, etc.

Essential Papers

Any authority looking at airline tickets, boarding passes and passports will examine names carefully. TSA and other security authorities often require that key papers (*e.g.*, airline tickets, passports) precisely reflect your full name. This applies to all persons in your travel group, *e.g.*, spouse, children (toddlers perhaps excepted). It starts by making sure that whoever books your cruise (and any associated airline tickets) accurately enters each full name on reservations and later-generated tickets.

Passports/Visas

Unless your ship's itinerary is confined to your home country (not often), you must prepare for a cruise as you would for any other international trip, to include passports, perhaps visas. Many countries to be visited may levy few or no visa requirements on day-visitors via cruise ship. But, check with the cruise line (through your agent if used) well ahead of time. Some lines will arrange needed visas for scheduled port visits, but also check specifically for visa requirements if you have an international flight itinerary.

- The cruise-line will often insist that your passport have more than six months before it expires as of the date your cruise starts. They may just be echoing requirements of countries the ship will visit, *e.g.*, many that require visas will not issue any under passports with less time.
- Lacking such a passport, or any needed visa, you risk being denied boarding on embarkation day...without refund or other compensation. You may also not be covered by travel insurance purchased.

Very occasionally, port officials in certain countries will require review of all passenger passports before clearing the ship for passengers to go ashore. If so, they often join the ship a few days in-advance, and the ship will announce a day or so before the port visit that the staff needs to gather all passports for inspection. Before you leave home, make machine or photo colour copies of at least the primary, facing pages of each passport...per details in the above linked article. Use the passports when instructed by authorities, *e.g.*, going through airport, airline or customs and immigration processing, processing for initial ship's embarkation. Take them with you on the rare occasion needed ashore per ship's announcements. Otherwise, once on-board, leave them in your cabin's safe and take the copies ashore instead, along with government-issued photo identification, *e.g.*, driver's license.

Boarding Passes and Tickets

Once booked, you need to promptly go to your cruise line's web site to "register". (Immigration authorities require that any ship leaving their jurisdiction have personal data of all passengers at least 3 or more days before cruise departure, but don't wait that long. Your agent or the line may have to mail these papers to you.) There (using your booking number), complete all details about all people cruising in your group. Data needed will include full names, addresses, phone numbers, social security numbers (or the equivalent for other countries), passport details, emergency contact names and phone numbers, how you want your shipboard account established/paid.

Once registered, the site will often allow you to print your boarding passes if within 60 days or so of the cruise. Others may mail them to you as noted above. You may get one set listing all persons in your group, or a set for each person. Each may also include the cruise contract and boarding instructions and times.

If you've paid for flights or airport-to-port transfers, you should also find vouchers or flight tickets (or Internet links to them for printing at home). If you don't have these key papers in-hand in some usable form at least three weeks before your cruise, notify your agent/cruise line immediately.

Each person will need his/her boarding pass and passport to be processed on-board.

Cruise-line Luggage Tags

Tags showing your name, cruise ship, cruise date/identity/number and cabin number are essential to ensure your large luggage reaches your cabin.

- Some lines will send you durable tags in-advance, to attach at least to your large luggage. If so, they'll often come with your boarding passes and vouchers for pre-paid transfers.
- Others will offer tags for you on-line. If so, print them...yes, they'll be on printer paper. Then fold each as indicated into a narrow strip, and reinforce it with transparent package sealing tape. When you're ready to attach one, wrap it around a luggage handle, information showing, and then staple multiple times or apply strong tape at the overlapping ends.

- If you make your own flight arrangements, you'll have to get yourself and all bags to the port/cruise terminal. You can wait to attach the cruise-line tags as you claim them at the end of your flight, but do so before your bags reach the port if possible. If you have none, porters at the terminal can create tags using your boarding pass data.
- If you have arranged services from the cruise-line to transfer your checked luggage from the end of your flight to the ship, you need to attach the cruise tags as you check the luggage to begin flying. Before you leave the air terminal at the port city's airport, get confirmation from any welcoming cruise-line staff about transfers and any responsibilities you have, *e.g.*, if flying internationally, you may have to claim your bags to go through customs and immigration processing before re-checking them into the cruise line luggage system.

Other Essential Papers

- If travel is international, take no more prescription medications than you'll need on your trip...with convincing documentation that they belong to you and are necessary, *e.g.*, labeled bottles with your name, copy of the doctor's prescription. If any contains a controlled/narcotic ingredient, make absolutely sure you will not violate any law of any country you'll enter...even as a through cruise or flight passenger, *e.g.*, as needed, have the country's written permission to carry the meds within its borders. Otherwise, the consequences in a few countries can be severe, *e.g.*, immediate confiscation, arrest, possible prison.
- Avoid paying duty on what you already own. See Proof of What You Already Own
- If you have purchased travel insurance, take at least a summary of the policy coverage and how to contact the insurer for help from wherever you will go on the trip.

FLYING TO/FROM PORT

Cruise ships sail from an increasing number of port cities. Most people must fly to get to them. If that applies to you, you have options. They include arranging your own flights (discussed later), choosing a cruise line fly/cruise package, choosing a cruise extension, or making a flight deviation request to suit your needs/desires to/from the cruise.

For all those options, if you miss your ship's departure (actually from any port), you are responsible for joining it at it's next port of call (can be very expensive unless covered by travel insurance and you are not at fault) or getting home on your own. Some sources imply that ships will delay departure for flights their line has arranged that arrive late. More accurately, they may delay, but only if it doesn't compromise the ship's ability to reach the next port on-schedule. Accordingly...

- If you're flying to (and from) an "in-country" departure port, insist that your flight is scheduled to arrive in plenty of time to reach the terminal...at least two hours before sailing. For return flights, see "Disembarkation" discussion below.
- If flying to an overseas departure port, get to the port city at least a day before cruise departure, even if it means you must pay there for an overnight stay, meals and transfers. Such "day early" arrivals are good industry practice and should be offered by your cruise line in any fly/cruise package. Your insistence on this should not be treated by the cruise line as a flight deviation request. If you'll have a long flight, you'll appreciate the chance to rest anyway.

A fly/cruise package means the line makes all arrangements for you to reach, go on and return from the cruise, including transfers...often at great convenience and confidence for first-time cruisers or those embarking at unfamiliar ports. Those packages have trade-offs...

- They usually cost more than arranging your own flight(s) and transfers..."usually" mentioned here because lines offering new, expanded or unusual "markets" may offer "free airfare". But take care about "free". If you see anything in trip descriptions or the total invoiced/quoted cruise costs that in-effect refunds/rebates for air support not used, the offer really isn't free...but it still may be desirable.
- The line's choices of flights/routes are often made semi-automatically, emphasizing cruise-line cost-avoidance (*e.g.*, using airline seats they've already blocked), then your cruise departure time. You have little control over their initial choices. Results occasionally have inconvenient flight times through busy air hubs with short connections, that require considerable walking, or that have quite long total times enroute.
- The line will notify you of proposed flight details at least 45-60 days ahead of the cruise. Promptly examine them. If any detail appears problematic, promptly question its wisdom (through your agent if used), and decline to accept until corrected or explained to your full satisfaction.

Most lines also offer cruise extensions to distant ports near or where cruises start or end. You can opt to spend 2-3 (perhaps more) nights at/near a port city immediately before and/or after your cruise. Somewhat pricey, they are integrated into your overall trip, with flight arrangements, quality lodging and transfers covered. Meals may not be included. They may include one or more tours. You shouldn't have to pay any flight deviation fee (below) with an extension. Cruise brochures, your travel agent and the cruise-line's web site will variously describe extensions available. If travel involves flying overseas, choose the number of days of any extension with intent to use at least some of the first day to rest.

Flight deviation requests. For an additional fee (*e.g.*, \$100+ per person), the cruise line will process (not book) your request for special flight arrangements. You place it through the agent that books your cruise if any...best made as you book the cruise, certainly long before its start. You can request dates, flight time windows or times, aircraft cabin class, even suggest an airline and flight numbers.

- The later your request, the more difficult it is to arrange. Somewhat economical seats blocked in-advance by the cruise line can sell-out rapidly. Late deviation requests can force the line to quote you costs to "cold purchase" tickets, at substantial extra cost to you, even using their wholesale or consolidator contacts.
- Your request may be one of hundreds being processed for that cruise, so make it simple, *e.g.*, clear objective(s) with few suggestions, alternatives or options...each requires some research by whomever helps you.
- The more specific or "exotic" your request (*e.g.*, non-stop flights, delays/stays enroute), the more expensive your flight arrangements will tend to be. Perhaps after negotiations, you'll be notified (through your agent if any) of the line's final offer of air costs (beyond the basic deviation processing fee) and itinerary details. As above, examine each detail for feasibility and whether it meets your objective(s).
- If you decline all offers, you still must pay the air deviation request fee.
- If you've booked a cruise with terms that basically say "airfare included", watch for unusually high added air charges, *e.g.*, just for using a different gateway airport. This may be just an effort to recoup some of cruise line's air costs.
- Once you're booked, if you have enough frequent-flier miles, and "your airline" or "alliance" gets chosen to transport you, you may be able to use your "miles"/points for seating upgrades. To find out, contact the airline directly.

Booking your own: If you're a seasoned traveler or have a good travel agent, you/your agent may do better by booking your own flights and lodging...better economy and/or flight dates/times/routes/seating and/or a hotel you prefer. However, If you must fly to one city/port for embarkation, and return from a different one, look at both options...self/agent-arranged flights and cruise-line fly/cruise packages.

- On most airlines, one-way domestic tickets cost significantly more than round trips for the same dates.
- With an international cruise, you/your agent face purchasing multiple one-way tickets for each traveler...often at much greater cost, *e.g.*, each can cost more than twice as much as a round-trip on the same dates. However, with some effort:
- You/your agent may be able to find "open jaw" tickets ([3]) for less.
- Your agent may know of budget airlines that offer one-way "econo" bookings, especially if you can be flexible about dates, routes, times enroute, on-board niceties...and occasional need to claim luggage from one carrier, take it to another terminal and recheck it on another airline while enroute.
- Cruise lines anticipate such costs/issues and block seats accordingly. If you ask early enough, your cruise-line may offer you a slightly better price than you caneasily obtain independently, perhaps including transfers between airports and ports.
- If you arrange anything on your own, keep any affected agent informed of your intent, efforts and results.

Regardless of how they are arranged, ensure you have completely adequate scheduled connection times to make flight check-ins, flight connections, the start of your cruise and return flights...to include extra time for unpredictable delays.

Consider everything that might make you late, *e.g.*, flight schedules, "tricky" connections, seasonal weather problems throughout your route, distance/time between any port and airport, ground transport dependability/availability, how far in-advance you need to check-in for flights. If you don't have fully-adequate time, choose other arrangements.

Such complexity, risk and cost again recommend that you use an agent and get quality trip insurance. And your trip insurer will expect "due diligence" from you in planning and booking transport to avoid travel problems.

Home-to-Port by Land

In contrast to flying, you might drive to your port city if practical without great effort and if the costs are right. With an adequate vehicle, you can take and bring home much more than allowed by air...quite useful for serious shoppers, family souvenirs and Christmas gift list items. Optionally, you can also visit the port area one or more days before or after the cruise. If your group is 3 or more adults or a family, you'll need a sizable vehicle for all, to assure comfort and room for luggage. If it looks tempting, examine and compare...

- Using a hotel/motel park-stay-cruise package near the port. Some let you park for 7-14 days at no or small charge (but no assurance of vehicle security) and offer shuttles to/from the port. Certain web sites specialize in finding such offers.

- Driving and parking at a commercial parking lot near or at the port. They usually offer shuttles to/from ships' terminals. They can be pricey for multi-week cruises, but do offer some security for your vehicle. Lots in/on the port are often much more expensive than "off-port".
- Renting a car/van one-way each way. This lets you "up-size" for comfort and capacity, and avoid parking costs, especially for long cruises. However, watch out for large drop-off fees, especially if the drive crosses borders or state lines. Check for rental brands that offer at least one office for vehicle drop-off and pick-up near the port, and offering shuttles to/from terminals.

If you can obtain quality, convenient bus or train service (as in Europe), you might obtain similar benefits, with simplicity and savings possible over a car. Examine the carrier's costs, reputation for punctuality, schedules, locations of terminals, transfers needed and any parking costs near home, and plan your departures accordingly.

Embarkation

This all starts as you reach your ship's terminal. The walking distance from ground transport to on-board ship can vary from 100-300 meters or so depending on terminal design and ship size. If anyone in your party has mobility challenges, request help in-advance.

If you are to board a large ship, examine the cruise line's instructions on when to arrive. Consider that:

- A large group will eagerly reach the terminal as early as they can...perhaps forgetting that the same staff (ship's, terminal and Customs) that processes them will first have to help nearly all on-board passengers disembark. If you are also that early, you'll likely spend considerable time idle in line waiting to be processed, with no guarantee of seating or shelter from the weather.
- Many other embarking passengers will precisely follow the cruise line advice and arrive at the designated time. That can create another line.
- Once all staff are ready to embark passengers, only very frequent customers of that line, or those who've booked suites, will receive any preferential processing.

If you have a choice, consider reaching the terminal 30-60 minutes after the time mentioned by the cruise line. If you're traveling with a personal group (*e.g.*, family), don't begin embarkation processing without all members present. In any event, reach the cruise terminal at least two hours before the ship is scheduled to sail. As above, if your agent or cruise line has arranged a flight that could cause you to reach the port later, seriously question its wisdom well in-advance. At the cruise terminal, give your large baggage (virtually no limit on numbers, but don't pack that much), with cruise tags attached, to porters for mandatory, separate security screening, then loading on the ship. (As above, if you have no tags to attach, the porters will help using details on your boarding pass.)

They deserve a modest tip. You won't see those large bags until they arrive at your cabin door...2-3 hours later, perhaps more. If you've paid the cruise line for airport-to-terminal transfers, and you have no intermediate customs processing, you may not see your checked luggage after initial airport check-in until it reaches the terminal, perhaps even at your cabin. But learn the full process and your responsibilities. As you start in-processing at the ship's terminal, officials will examine your boarding passes and passports. Then:

- You and your hand-carried items will be scanned, much as at an airport...but without limits on numbers or non-alcoholic liquids.

Note about liquor: Nearly all cruise lines forbid bringing "personal liquor" on-board. Indeed, among some cruisers, smuggling it has become an art form. The above scanning will look for it, as will scanning of your large luggage. Any found will be "kept" by the ship until you disembark. If you've put any in your large luggage, and the bag hasn't reached your cabin within a reasonable time, check with the Pursor; it may be "on luggage hold" on a lower deck for you to personally claim...sans the liquor. Modest quantities of wine may be allowed, and hand-carried soft-drinks seldom get refused. Check the cruise line's policy before packing.

- You'll often be asked to fill-out and sign a current health statement...one for each person in your group. If anyone reports or exhibits symptoms of something that might be communicable, he/she may be interviewed. At worst, he/she may be denied boarding, at no liability to the cruise line. Law requires them to minimize health risks to all the (perhaps) thousands of close-proximity passengers and crew on-board. This is another issue usually covered by good trip insurance.
- You'll be shown to a processing counter to provide identification, set up a shipboard charge account and be issued a cabin key card for each person in your group. (See "Ashore" and "Buy" below)
- From there, you'll begin boarding. On the way, you'll each have an electronic identification photo taken...computer-linked to your cabin key/card. After that, you are free to walk about on-board. Later, as you disembark and re-board, you'll have to use your card, and the photos will electronically pop-up on a monitor used by ship's security staff.

- On the way, you'll often encounter a ship's commercial photographer. All pictures they take...throughout the cruise...have no cost unless you choose to buy one as you find it later in the ship's photo gallery. They tend to be rather pricey.

Once on-board, the buffet and casual food counters await...usually on upper decks. You'll often be instructed not to go to your cabin until 2 PM or so (your cabin steward is completely cleaning, sanitizing, and changing all linens and towels in your cabin and many others). So, after food if any, it's a good time to walk about the ship to get oriented.

- Top-side you'll probably see one or more swimming pools, other sports and exercise facilities, bars, perhaps a spa and a lounge.
- Likely on lower, public decks, you'll find the Pursor's Desk (aka Customer Service), often a concierge, shore tour ticket office, future cruise sales desk, shops (not open), and somewhere an Internet cafe.

Before sailing, there will be a safety at sea briefing that everyone must attend. Cruise lines and captains take this maritime law requirement seriously. You'll find instructions in your cabin, and papers and announcements will tell where and when to go. It includes learning the location of your emergency "muster station", ways to get there from your cabin, emergency signals and procedures, and how to wear your life vest. You may or may not need to take or don your life-vest...stored in your cabin. All ship's services will be closed during this time. Truants may be called to a separate briefing at staff's convenience...with evaders possibly jeopardizing permission to stay aboard. If anyone in your group has mobility problems, this is a good time to tell the staff serving your muster station, so they can pre-arrange special help for emergencies. If time permits after the briefing and before sailing, meet your cabin steward and discuss any needs or preferences, *e.g.*, ice for the cabin at certain times. Then go top-side for departure...always interesting, often scenic (take your camera if light will be adequate); a bon voyage party is likely.

GET AROUND

The key advantage of a cruise ship is that it does most of the "getting around" for you. See "Understand (above) and the following sections for details about ships and port visits. Basically, you unpack once and then visit the ports on your ship's itinerary...only repacking at the end of cruise. As you explore the ship, you'll note that instead of numbers, decks (not floors) may have fanciful names. You may find yourself referring frequently to the maps in the elevator and stairwell areas to figure out whether the Lido Deck is above or below the Promenade Deck. The biggest ships can be 15 or more decks deep (counting bars and whatnot perched above the pools), making even the most conscientious stair-climbers resort to elevators from time to time. Thankfully, elevators will have an outside and inside list of facilities for each deck. But they often do not indicate if any is aft or forward...back to the maps. Only rarely does a ship fail to visit a scheduled port. This most often stems from adverse weather. If such threatens, the captain will avoid the effects of the weather as much as possible, and will make announcements explaining what is happening and if you'll visit alternate ports instead.

Fig. Cruising Southeastern Alaska's Inside Passage.

Some ships have been outfitted with millions of dollars worth of art and elaborate interior decor, but generally after a few days there isn't that much to see on most cruise ships. The real sights are ashore. Some ships travel to geographically interesting areas such as Alaska or Scandinavia where they make detours to view fjords and glaciers up-close. Generally speaking, the smaller the ship, the better proximity to scenery you can expect, because they won't need to stick to deep and open water. On large ships, other scenery may be too far off to really enjoy its details, though binoculars help. Depending on the region and season, you may spot whales, dolphins, or flying fish swimming nearby or even following alongside. Lacking those benefits, the real sightseeing opportunities come as you approach and reach port, and as you take shore outings discussed below.

Do

As you plan each day's activities, everything on-board will be based on ship's time. Depending on it's itinerary, the ship will usually change its time to agree with any time zone it has entered. This assures that you can take advantage of all activities and tours, on-board and ashore, with confidence about time. Your cabin phone will follow that change, so feel free to use its wake-up call capability to ensure you miss no event.

Aboard

Large ships will have most or all of the features discussed below...mega-ships even more. Smaller ships (*e.g.*, 600-1200 passenger capacity) will have many of them, but in fewer numbers or smaller scale.

You'll be surrounded by water you can't swim in (it's passing by at 15-20 knots or so), but all but the smallest ships will have at least one "swimming pool" (perhaps covered, otherwise usable only in warm climes) and deck chairs. The pools won't be great for swimming laps, but some new ships are being equipped with small, swim-against-the-current pools. Most are filled with processed seawater. Without the legal restrictions imposed on land-based facilities, most cruise ships have a casino except for Disney. Expect more emphasis on gaming on ships catering to Americans than for Europeans. Don't expect table games or machines with payoff rates even close to those found in better land-based casinos; concessionaires must pay dearly for the space. It will be open for gaming only while at sea.

However, on larger ships, Las Vegas and Broadway are the models for entertainment. They'll variously feature singing-and-dancing shows, feature singers, comedians, magicians, and other live entertainment. Shows typically follow dinner, but may precede it for those who opt for "late" dinner seating. During and after shows, other venues offer small bands, piano bars, and dancing to live music or a disc jockey. A movie theater is found on most ships, playing movies similar to those found on airlines. There is usually a library on board for your reading pleasure but don't expect the latest novels unless left behind from an earlier cruise. If cabins have DVD players, the library may have a modest collection of titles. It may offer electronic or board games to check out. Quite commonly, you'll find an Internet cafe (discussed later). Topside you should find someone issuing equipment for basketball, table tennis and other uses.

The ship will often offer space and seating to support impromptu or organized bridge, even tournaments. And staff very often have trivia and other contests. And on most sea days, you'll see at least one large Bingo session offered. Shopping is readily available, with shops on board. (See "Buy" below.) They'll be duty free but don't expect big bargains. They too will be open only while at sea. You'll receive a daily newsletter with a schedule of activities...apt to mention art auctions (reportedly the "most dangerous place on a ship"), bingo, kitchen tours,port and shopping lectures, cruise enhancement lectures (by naturalists, historians, political scientists, et al), arts and crafts lessons, poolside contests, dancing classes, etc. Family-oriented cruises will have many age-specific activities geared to the kids and teens.

Most ships have a gym or health center with exercise machines. They often offer instruction programmes in exercise regimes or Tai Chi, etc., some at small extra cost. Many people use the "promenade" deck or topside track. The former usually loops around the ship around mid-decks – for walking or jogging, but may have stairs that interrupt you. If so, a topside track might be better if available. Some ships find room for putting greens, golf simulator, a basketball or tennis court (enclosed by ball-catching nets) topside. Some very-large ships have ice rinks, rock climbing walls, "surf parks" and other activities. Spa facilities are a staple of cruise ships. Everything from massages to hairdressing to exotic health and beauty treatments are available...at substantial extra costs.

Look for laundromats in ship descriptions...not all ships have them. If the cruise spans seven or more days and the ship does not offer them, you may have to pack substantially more clothes. Any facility offered will have two or more pairs of washers/dryers, detergent and softener dispensers (most taking coins/tokens), and irons and ironing boards. All ships offer laundry services, but they'll be pricey. Lacking laundromats, some folks bring detergent (*e.g.*, Woolite) to hand wash select items in their cabin...though humidity creates a drying challenge.

Ashore

You must use your cabin key card every time you leave and re-board the ship. It's how they determine if you're aboard...and how ship's security staff recognize you. Otherwise, they might leave you behind.

- You'll regularly be instructed to return to the ship at least 15-30 minutes before its scheduled departure.
- Near departure time, based on computer tracking of who's not back, you may hear shipboard announcements for passengers to contact the Pursor's desk...they're "missing".

Research in-advance each place you'll visit. That can greatly enhance your cruise experience...at little or no cost. Many ports and nearby sights are covered by Wikitravel pages, travel web sites and books.

The ship will usually dock at a pier. If none is available, it will anchor or moor off-shore, and ship's lifeboats or commercial boats will tender you to a convenient place on-shore. Your detailed itinerary will indicate how your ship will visit each port.

- If you're on a large ship, this tendering may take some time. Ship's tour members usually go first, then "early-birds" who've obtained first-come-first-served tender tickets or numbers. Unless you've purchased a ship's tour, plan your time ashore accordingly.
- If weather generates rough seas at any port where your ship anchors/ moors, tenders to shore may be cancelled. The line may offer reimbursement for the disappointment, but will refund the cost of ship's tours purchased but unavailable.

Ship port visit times usually allow passengers to go ashore by 7-8 AM, with ship departure often at 5-6 PM. Earlier or later departure times can be affected by tides, distance to next port or special tour needs, while disembarkation may be delayed slightly by port customs clearance of the ship or passengers. At special stops, some ships may stay later, perhaps overnight.

The shore excursions office will offer a variety of sightseeing tours, cultural visits and organized activities (*e.g.*, scuba, snorkeling, kayaking, bicycling)...offerings dictated by the nature of each port, its climate, time of year and time in port.

- Ship's tours at major ports often employ large air-conditioned buses for 30-50 people. Popular ports will also have large numbers of air-conditioned taxis...sedans able to carry 3-4 adults, and vans for 6 or more.
- All-day tours can fill virtually all your time in port. You'll usually find half-day tours offered as well, morning and afternoon...worth considering if a walk-about or shopping before or after lunch also appeals.
- If you are a confident traveler or you've visited the port before and want to tour, you may do well hiring a car with driver-guide to take a tour of your choice and design. It can involve just your family or 2-4 fellow passengers or so.

Ship's tours often cost substantially more than equivalent tours well-negotiated directly with locals, *e.g.*, you may hire a taxi or van with driver/guide at $40-50 per hour for 4-6 people. A half-day ship's tour can range from $50-100 per person, with whole-days $125-200+. But for that extra cost, ship's tours provide:

- Great convenience compared to finding things to do and making arrangements yourself after you go ashore in a strange place.
- Some confidence if you will venture far from port.
- Some assurance in especially "entrepreneurial" locations that you won't be scammed by a taxi driver.
- Greater assurance that the ship will wait if your tour is not back on time.
- Again, it may wait for you if you are on your own, but only if it can reach the next port on-schedule without compromise or high added cost.

Very popular ship's shore excursions may fill-up days before you reach port – some even before you set sail. You may be wise to research them and alternatives and sign-up online well in-advance as justified, *e.g.*, if...

- You have your heart set on swimming with dolphins or stingrays, or climbing a glacier.
- You're focused on a full-day tour to a truly special locale some distance from the port.

At some ports, the ship may have to dock among commercial freight operations. Walking from the ship to things worthwhile may be through unpredictably dangerous activity. Look at the ship's newsletter for port conditions and listen for port adviser announcements. At such docks, most ships will arrange a shuttle from the ship, to a terminal (with taxis, etc.) or to a shopping area or downtown. If so, also look for details about a return shuttle. Lacking any shuttle, you should request transport that avoids danger, even if at your expense.

The UV content of the sun can be very high on any tour, especially on or near water at any latitude during the Summer. See "Stay healthy" below..

BUY

Money Matters

Cabin key. For both convenience and to foster a casual-spending atmosphere, most cruise ships run a "cashless" system in which you use your cabin key (card) tocharge all on-ship expenses, sometimes except for gambling. Two or more cabin keys/cards can reference one credit card or cash account (deposit given in-advance), *e.g.*, for couples and families. At the end of each cruise segment, the ship will use your credit card or cash deposit to settle the final balance of your account.

- You can obtain the current balance of your account, with a list of all charges, at any time, sometimes even on your cabin TV. You should review those details at least near the end of cruise to allow you time to question any charge.
- If linked to your credit card, your account balance will be automatically charged to it; many lines charge your credit card with expenses during the cruise. At cruise end, all is settled automatically...allowing use of your cabin key at least through the last evening.
- You must personally settle any cash account. This risks long lines at the Pursor's Desk on the last full day of the cruise as many others also pay their accounts. Once paid, you may not be able to make any more charges before disembarking.
- At debarkation, anyone who has not fully settled their account will not be allowed to leave.

Tipping

Tips on-board take two forms...the surcharges for special drinks, specialty restaurants and some services (discussed below), and (ultimately optional) tips for the ship's staff (not officers) usually levied at the end of the cruise. Learn the cruise line's recommendations for tipping the ship's staff. They suggest "appropriate" optional amounts...a crucial part of staff income. Amounts recommended may vary somewhat by your class of cabin.

- Many cruise lines add those amounts to passengers' shipboard accounts automatically. This method ensures that all staff contributing to your cruise experience (*e.g.*, cooks, cleaning and maintenance workers), not just your waiter and cabin steward, receive some extra income and recognition.

- No automatic gratuities go to ship's officers.
- You can opt to decline or adjust this arrangement as desired before cruise end. Royal Caribbean and some others offer the option in advance.
- As inducement to regular customers, some cruise agencies may pay standard gratuities.
- Whether automatically or in-person, any crew member who has especially enhanced your cruise experience deserves something...perhaps in-addition to automatic amounts.

Some cruise lines have a "no-tipping" policy, often aimed at the European market where tipping is sometimes alien and can frighten customers away. In reality, gratuities are often built into the price of those cruises, on which passengers usually also pay taxes.

Shopping On-board

Cruise ships take advantage of their international/at-sea status to sell a variety of duty-free items (*e.g.*, liquor, jewelry (costume and precious), cosmetics, perfumes) at decent if not outstanding prices. Large ships usually have boutiques selling logo clothing/souvenirs, perhaps offering casual and evening wear (a few even tux rentals). Others shops offer basic sundries, candy and over-the-counter drugs. Shops will only be open while the ship is at sea.

- You may see "sidewalk sales", specials on jewelry and watches, and towards the end perhaps a clearance sale on outdated logo and other items.
- All bottled liquor purchases on-board will be held until the last full day of your cruise, and then delivered to your cabin...likely well-boxed, needing just a bit more padding in the box to be ready for carrying or (after plastic bagging) to go in checked luggage to fly home.
- Most ship's merchandise is subject to duty if all of your trip purchases exceed your duty exemption as you return home. For expensive items (*e.g.*, jewelry), check if they were made in your home country. If so, with a proper certificate, you may not have to count them against your duty exemption as you return home.

Shopping Ashore

Shopping remains a highly-popular activity, with most ports offering at least handicrafts and souvenirs reflecting the destination. Others offer wide ranges of merchandise, to include clothing, electronics, jewelry and consumables. A crude rule of thumb: the farther you venture from the dock area, the lower the prices may be for given types of items.

Fig. Cruise's Pedestrianised Street, Limerick.

If there is a chance that you'll use your credit or debit card:

- Let the issuing bank know in-advance where you'll be traveling.
- Follow good practices for ensuring it won't be lost, and won't be misused by others.

Cruises to certain destinations may stop at ports offering serious shopping...duty-free and/or highly competitive, *e.g.*, Saint Martin and Saint Thomas in the Caribbean, and (perhaps except for "designer stores") Hong Kong and Singapore. For example discussion of duty-free shopping and customs obligations, see Saint Thomas#Buy. If inclined to seriously shop somewhere not familiar to you, do some on-line research about what to expect, stores, tax/duty, and what to watch out for.

The ship may have a port shopping adviser who can provide useful information and may recommend certain merchants. Those advisers are often employed by the cruise line or merchandising firms that receive very large fees from the merchants they recommend.

This does not impeach the reputation of those merchants, nor should it question the worthiness of their competitors. But it may increase the final prices that touted merchants demand. In desirable shopping locales, most merchants are quite responsible.

At other locations, take care with final costs beyond labeled or negotiated prices, *e.g.*, duty, or tax such as Europe's VAT not reflected in prices...often 17-25 per cent. Though eligible for them, you may be able to obtain refunds only at certain places, *e.g.*, airports as you leave the customs jurisdiction to return home, and only for each purchase that exceeds a substantial amount. The last cruise port for a ship exiting such jurisdictions may have no "standing", nor the port an office offering or processing refunds. Getting refunds by applying by mail after you are home will be problematic. Such challenges cause serious shoppers to seek-out cruises to and wait to reach truly duty/tax-free and highly-competitive locales. Wherever you shop, know what you're getting and what it costs elsewhere.

Eat

Nearly always, meals are included in the price of the cruise. This includes poolside snack bars where you can order a burger or something and walk off without paying. (It's not "free", of course; you paid for it when you bought your ticket.) On virtually all cruise ships, you'll find a buffet...usually on one of the upper decks...available during all meal times and usually offering something from early morning to late evening...again, free. On better ships, buffets can seem almost like pure extravagance. Room service is usually available at all times, except after 1:00 AM the night prior to disembarkation on most cruise lines. Sometimes, you may have to pay a gratuity or late night charges when applicable.

At normal meal times, you'll find seated dining with full waiter service available, usually with a multi-course menu featuring variably fancy dishes. Most of the menu will change every night. At least one dining room will offer seated breakfast and lunch, seldom with a pre-assigned table. It will be open for 2-3 hours around the usual time for any meal. Note: Damp swimwear can damage dining room chairs; don't wear it there, even under a wrap or cover, even if quite dry. Traditional evening dining service is at pre-set times...usually early/main and late sittings. If you choose a time (and table size) (best when you book your cruise), you'll be seated at the same table at that time every evening.

- If your table preferences haven't been met (first indicated on your cabin key card), contact the maitre de on-board for help as soon as possible after embarkation.
- Exceptions to your usual time and table may may occur when the ship is in port and many passengers are eating ashore. (The daily newsletter should mention this.) If so, you may share a different table with "strangers", even in a different dining room.

Table size can vary from 2 to 8-10 people, perhaps more. Round tables for 6-8 seem conducive to easy conversation among all. Unless you/your group fully occupies a table, you'll meet other guests as table mates...usually an interesting time, with repartee beyond what's possible with a new set of strangers at "open sitting". It also allows your waiters to learn and anticipate important needs and preferences, *e.g.*, kosher, vegetarian, food allergies, drink preferences and timing. To assure well-timed service, reach your table within 30 minutes of when the dining room opens for your sitting. Large kitchens must serve several hundred (perhaps thousands of) passengers at each sitting, one course at a time, with expectations of freshness and proper temperatures.

In recent years, to respond to some guests' dislike for scheduled dining, cruise lines have introduced freestyle, choice or open seating options which allow dining at any time during dinner hours. This may be offered in separate dining rooms. Menus will be the same as for scheduled sittings. As you enter, you may have to wait for a table, just as you would without reservations at home. If it's especially busy, you have just a few in your party, and are willing, tell the Maitre de that you "will share". It greatly helps him/her efficiently use tables/seating, and may speed getting you seated. Most ships also offer specialty restaurants, often with international themes...usually by reservation only. (If you have no reservation, you can try "will share", but don't rely on it; most diners with reservations don't expect to share.) Some of them have surcharges (*e.g.*, $20-30+ per person) for exceptional service and dishes...most well-deserved. If you normally dine at fixed sitting but plan to use a specialty restaurant any evening, tell your regular waiter the evening before. Recommended dinner dress in dining rooms varies somewhat across cruise lines. For details, each cruise-line explains its expected dress code on its web-site, and you should find details on web sites for cruising.

- Many "main line" ships declare dining for most evenings as requiring just "country club" or "resort" casual wear...collared shirt and slacks for men, sundresses or nice blouses for women.
- They'll also often have one or two "formal" nights per week...tux or dark suit and tie for men, evening wear for women.
- Some luxury lines may declare every evening as semi-formal...suit and tie for men, evening wear for women.
- For any evening, the same dress standards apply in specialty restaurants.

You'll embarrass yourself, your table mates, your waiter and others if you go to your dining room for dinner wearing jeans, shorts, a tank-top, or similar casual/pool-wear. You may also be denied seating. If you have no interest in "seated dining" on formal nights (or any evening), you can use the buffet for dinner instead. Food offerings will be somewhat similar to that in dining rooms that night, perhaps lacking items that require complex service. The buffet offers another benefit:

- If you must fly to/from port, and may too-easily exceed your luggage limits, *e.g.*, must pack for small children and yourself, have a long cruise on a ship with no laundromat...
- Consider leaving semi- and formal-wear (and related shoes/accessories) at home. You can reduce packing space otherwise needed by perhaps 25-30 percent, and weight similarly.

DRINK

Typical staples such as coffee, tea/iced tea, lemonade, juices (at breakfast) and iced tea are available at no charge. Other drinks are usually not included in the cruise price, even if the cruise's promotional brochure says or implies "all-inclusive".

- Those "others" (*e.g.*, soft drinks, bottled waters, fresh-squeezed juices, specialty coffees, beers, wine, mixed drinks) can be pricey. Prices listed will not include a 15-18 percent service fee/surcharge.

- Expect to find one or more well-stocked bars on all public decks, catering expertly to your preferences, many opening mid to late morning and some staying open very late.
- All dining rooms will have a very good wine list, with a few offerings by the glass, as well as bottled waters, mixed drinks and specialty coffees.

Many cruise lines offer drink packages for unlimited soft drink refills, some for specialty coffees, wine, even a few for mixed drinks, but each for a substantial additional cost that often includes the service fee.

Understand the terms clearly before committing to any. Examples:

- A "pop/soda pass" often runs $30-45 for a week, which can equal 3-5 separately purchased drinks every day. (A considerable average daily consumption rate.)
- A recently-reported mixed-drink package runs over $300 per person for a week, while most separately-priced drinks run $7-11 or so each. Again, this means a considerable daily consumption rate to "break even".

Not counting the service fee, some mixed drinks may be cheaper than you might pay at better watering holes ashore, but don't count on it. However, you'll usually find drink specials mentioned in the ship's daily newsletter. These costs prompt some people to try to bring their own. But most lines forbid bringing liquor on board, and any found (at embarkation or as you board from later port visits) will be "held for you", and returned on the last full day of the cruise in the same packaging as received. Some lines will allow you to initially bring 1-2 bottles of wine per cabin, and most will allow you to bring your own soft drinks. Some ships are primarily party vessels, full of young adults taking advantage of duty free alcohol and (perhaps) lower drinking ages in international waters. You may identify them by their extremely uneventful itineraries: straight out to sea, stay there for much of the trip, then back to port. Their advertising is usually also not particularly subtle. If you want one, you'll recognize the signs; if you want to avoid one, likewise.

Mainstream cruise lines avoid unbridled drinking by:

- Requiring at least one occupant of each cabin to be a minimum age (with some exceptions for legitimate families) and/or by not serving alcohol to anyone under 21.
- Training and directing wait staffs to control service to minors and those who've had too much.

SLEEP

Let your travel agent know your cabin preferences or needs in advance; *e.g.*, to ensure your cabin assignment meets all needs. Shortly after arriving at your cabin, introduce yourself to your cabin steward and discuss any preferences or needs for service, such as ice at certain times, softer or firmer pillows, etc. They will often have an assistant and will both be working together as a team. If you brought sanitizing wipes or solution, you might wish to sanitize key surfaces like the lavatory counters, telephone handset, TV remote, and desktop. For families, in addition to the bedding noted earlier, some cabins will have a pull-down bunk-bed (not appropriate for small children due to safety, or full-sized adults due to length and weight) and/or a pull-out sleeper-settee or sleeper-sofa.

- If you've booked as a family, your steward will unlock any "pull-down", for your use whenever desired. It will be made-up each morning and prepared each evening.
- If there are more than 2 adults, your cabin steward will prepare the sleeper-sofa each morning and evening; even a second bed if included with the cabin.
- Suites often have more options, with the cabin steward (or butler) ready to explain use of all and prepare as needed.

If your cabin is not as desired or needed on arrival, resolve it through your cabin steward before you retire the first night.

STAY SAFE

Because of the numerous advancements in modern shipbuilding and other technology, cruise travel is generally very safe. With the exception of some of the smallest lines (which may have a very limited staff), the crew of your ship are all well-trained to calmly and efficiently handle any emergencies that may arise during the cruise. Occasionally, even some considerable problems may be discovered and repaired while going completely unnoticed by the passengers.

Aboard

In the event of a life-threatening emergency, instructions will be given on where to report (most likely your lifeboat station, designated at the pre-departure safety briefing) and what to bring with you (probably just your life vest and cruise ID card). Remain as orderly as you can and do not panic as you are evacuated; although cliché, acting out of place will only make matters more complicated and increase your chances of injury.

Remember to be aware of your surroundings at all times, especially at night. You shouldn't feel worried about going about on your own, but should be extra vigilant around bars where running into intoxicated passengers is a possibility and any altercation could escalate into a very dangerous situation. Keep an eye on your belongings and don't flaunt nor take them with you everywhere; leaving your iPod on a pool chair unattended while you quickly order a drink is just asking for someone to snatch it from you. Lock expensive items and jewelry in your cabin safe, then use or wear it only when appropriate.

If you win a substantial amount of money in a ship's casino (unlikely, but not impossible), ask to have your earnings credited to your cruise account or given to you in the form of a check before disembarking, not in cash form for your wallet.

Families traveling with children should be cautious as well. While family-friendly lines like Disney are "age-proofed" for their safety, other lines and older ships typically are not and there are many hazards that could put them in a dangerous situation; *e.g.* being left unsupervised on a balcony, falling down the stairs, swimming in a pool without a lifeguard, etc. Older teenagers should be given some freedom about what they'd like to do and where they want to relax during the cruise, but you should always know the whereabouts of younger children. Always report anything suspicious or concerning to a crew member; they will investigate or get someone who can investigate the matter. Trust your instincts as well. If a certain situation or activity in an area of the ship doesn't feel comfortable to you, it may not be.

STAY HEALTHY

Sunburn

As noted earlier, too much sun exposure (on-board or ashore) ruins cruises for more people than any other cause. Sunburns (and their pain) can last until well after you return home. For a few, they may later induce cancer or other permanent skin damage. Rather, before you'll be exposed to sun for more than a short time, take effective precautions with clothing and sunblock so that you can fully enjoy whatever recreation you choose. The ship's store will offer sunblock but at highly premium price, and a few kinds of hats; stores at ports may offer similar products, but almost always at prices higher than at home. (See also Sunburn and sun protection.)

Queasiness

Some people experience queasiness on cruise ships. This is very unlikely on large or recently-built vessels, which consistently have highly effective stabilizers. Even so, some sensitive inner ears may react to even imperceptibly slow and gentle rocking of a calm sea, while small, intense storms can make the ocean rough for a day or so.

- If you know it may be a problem, pick a cabin location that minimizes movement when the ship rolls and pitches; *e.g.*, amidship and/or on a lower deck.
- Over-the-counter motion sickness medications (*e.g.*, Dramamine, Bonine) usually help; though reportedly, Draminine has a tendency to cause drowsiness that can make it impractical to use for the duration of a cruise.

- Prescription trans-dermal Scopalomine patches can be very effective. But some people have troubling side-effects, so test a patch before the cruise if you like the convenience they promise.
- Some people find relief with special wristbands that claim to stimulate pressure points that are believed to counteract the nausea of motion sickness; their effectiveness has not yet been independently verified, but may be worth considering.

Staying well can be as simple as eating (and drinking) responsibly. Generous helpings of that beautiful, fresh pineapple for breakfast every morning can cause problems, as can the portions of wonderful and rich delights at dinner. Marine biologists know that the calf of a blue whale can gain up to 30 pounds per day; the next fastest weight-gaining mammal may be the cruise passenger - actually known to gain 6-7 pounds per week if he or she "over celebrates".

Outbreaks

Passengers and crew are susceptible to communicable diseases (*e.g.*, the flu, colds, Norovirus), but thankfully this happens only very occasionally due to great effort and care by the ship's staff, and by passengers who cooperate with health protections.

It occurs because large numbers of people from countless places have close proximity, share facilities, and forget to be responsible. Prevention: Most maladies spread much in the same ways as the common cold. Stifling coughs and sneezes into your sleeve helps greatly. And ships regularly provide hand disinfectant dispensers at entrances to dining areas; use them, but don't rely completely on them.

You can help yourself if you:

- Wash hands very frequently and thoroughly.
- Substitute "elbow bumps" for handshakes with staff and fellow passengers; at the very least, wash up well after shaking someone's hand.
- Use sanitizing wipes or solution on key parts of your cabin occasionally. This is no criticism of your cabin steward, who's efforts will also be thorough.
- Use your cabin bathroom rather than public restrooms whenever you can.

If you become ill: Report it as soon as possible, very preferably by cabin phone.

- The medical staff can help considerably, may attend to you in your cabin, and may direct you to remain there and cancel any port visits until you recover (usually within a couple days) to avoid spreading the malady. Such direction is often reinforced by law in ports.

- If the illness is deemed the ship's responsibility (*e.g.*, several other passengers have or have had it), you may not be charged for medical services and may receive reimbursement for missed ship's tours.
- Specially-equipped teams of ship's staff will thoroughly clean and sanitize any "accidents" in your cabin. So report them, and report public accidents as well.

Ships that see "possible signs" of an outbreak (even several passengers with sea-sickness) may set up extensive precautions and health/sanitation protocols to limit spread of any infectious agent. This can include hand sanitizer dispensers at entry to all public areas, barriers to self-help in buffets, nearly constant sanitizing of railings, door handles and public restrooms everywhere. Make use of and respect those measures.

Food/Drink Ashore

Although shipboard food and water will be sanitary, the usual precautions for overseas travel should be taken when eating and drinking ashore.

- Advance research about your port visits should include looking for possible health risks.
- In well-developed countries and "touristy" areas, food and water may be safe. In addition to research, consult the ship's port descriptions and the port adviser. Even with assurances, take care with sidewalk food stands and road-side cafes. Crowds of locals only indicate popularity and low cost, not necessarily safety.
- In lesser developed countries/ports/areas, take care to avoid local water and food-borne bacteria or parasites, including drinks made with ice a bar/restaurant may make using local tap water.
- If in doubt about getting water ashore, buy it in sealed, labeled bottles.
- Restrooms in undeveloped or rustic locations may not have bathroom tissue. Bring a small pack of tissues with you anytime you will be exploring outside of the port.

Carrying Bottled Water

Few ports prohibit bringing sealed bottles of water ashore. You'll see pricey offerings each time you leave the ship. Ask a ship's officer in-advance if sealed bottles are necessary; if not, you could (re)fill your own, with buffet beverage/water dispensers perhaps offering better tasting water than that in your cabin.

Medical Staff

Nearly all cruise ships have one, with an accredited doctor. Larger ships may have two or more nurses. Most will offer open hours ("sick call") in the morning and late afternoon for routine ailments, with on-demand response (even in-cabin) for injuries or major illnesses. Most can effectively render first- and second-aid, perhaps more depending on the doctor's experience and on-board facilities, *e.g.*, X-ray. They carry basic medications and supplies typically needed for cruises. Don't depend on them to replace medications you must use, even with near-equivalents.

- If the ship's itinerary is quite repetitive, the doctor may have standing with pharmacies in some ports. They may be able to issue usable scripts for your needs, but the ship's staff often must pick up the medications.

Unless your problem appears to have been caused by some condition or event that's the ship's responsibility, expect to be charged a fee for their services. Your regular health insurance may not cover such fees, but good trip insurance usually does. Some ships have modest dialysis capability. If you need it, learn in advance if it's the type you need and can reserve time slots. If available and as needed, reserve the service/times through your agent if you used one. Check with your health insurance agent to see if your regular policy covers any or all of what will be a substantial cost. Though the doctor and staff often wear ship's uniforms, many are concessionaires, not ship or cruise line officers. Nevertheless, for serious conditions, the doctor's decision about continuing your cruise under illness or injury will usually prevail. Medical evacuation can be unbelievably expensive, and so should be covered by full-spectrum trip insurance.

Physical Limitations

If you have physical limitations, have your agent arrange needed services in advance, *e.g.*, wheelchair to embark/disembark, for port visits/during the cruise, perhaps even to rent a "power chair" motorized wheelchair (usually only available for round-trip cruises).

CONTACT

Many cruise ships are now equipped with cell phone-to-satellite transponders, which take over automatically at sea to provide wireless phone service to the world throughout the ship. Except for satellite delay, your cell phone works just like at home and bills its usage back to your regular cell phone bill.

- Be aware that as your phone switches over to that "Cellular at Sea", you are roaming at the ship's pricey per minute rates; the allotment of minutes on your plan doesn't apply here. Data should work as well, but cost can add up fast if your phone automatically checks e-mail.
- Once the ship docks at a port with cell service compatible with your phone, and you then turn it on, your phone will likely use shore signals instead.
- Experienced cruisers wanting to avoid ship's costs simply never turn on their phones while on-board.

Most ships offer ship-to-shore phone service from your cabin, but again at rather expensive rates. They may also levy heavy tolls on people at home who contact you by phone on the ship. On-board Internet cafes and Wi-Fi hotspots are increasingly common, but the rates also tend to be fairly steep and the speeds (relying on high-latency satellite uplinks shared with ship's business) can be unimpressive.

- To "save" some money, consider buying an adequate block of minutes, at a lower per-minute rate, early in your cruise.
- Wireless coverage points may limit where on board you can use your laptop. (Remember that most of the ship is constructed of steel...a particularly unfriendly material for dependable wireless connections.)
- Some new ships have wired or wireless networking available in cabins, perhaps for another fee.
- Avoid depending on a ship's service to download major files of any kind, *e.g.*, video clips, software updates. Interactive networked gaming will be a joke. On-board with your own laptop, consider using only e-mail...after you turn-off all settings that allow automated updates to software, to include (heaven forbid) anti-virus/firewall suites. Turn them back on and wait to complete updates if using your laptop and faster services ashore, and promptly back home.
- Avoid depending on the day before disembarkation to conduct essential business. It's usually the busiest day for on-board internet use, with unusually slow response because passengers are checking flight details, using up surplus pre-purchased time and doing other tasks, all while ship's business is especially intense.

Internet ashore If you find these cautions worrisome, you can usually find Internet cafes at or near many ports...often offering much better rates and speed. Ask ship's staff for their personal experience and advice; many of them use those port cafes as well. Anytime you use any computer while traveling, ensure that all private/sensitive portions of your sessions and data are secure, *e.g.*, avoid doing financial or highly-personal business, use your own laptop if possible, have pro-active security capabilities installed and running, use wired Ethernet instead of WiFi if possible, set up a unique/complex password just for the trip, look for the https in networked sessions anytime privacy or security is needed. See more discussion at Internet access.

GET OUT

This is a recap of a typical disembarkation process. Near cruise-end, the Cruise Director will give a briefing that covers specific details for that ship/cruise and debarkation port, to include possible local customs and immigration (C and I) processing. Called "debarkation" or "disembarkation", this involves getting perhaps a few thousand passengers off the ship as efficiently as possible. You can't all leave at once.You'll likely receive a questionaire several days before cruise-end asking what travel arrangements you have made to return home. Your answers (*e.g.*, scheduled flight departure date/time) will determine in what group you will disembark.

- Two or so days before cruise end, you'll receive luggage tags indicating your debarkation group. Instructions that come with them will indicate the time your group is scheduled to be called to leave the ship.
- If your group's scheduled disembarkation provides too little time to comfortably reach a necessary place on-time (*e.g.*, your airline's check-in counter), notify the Pursor's Office immediately.
- On most ships, those tags will be colored and/or numbered. You can obtain more from the pursor if needed.
- On the last night, place a tag on the handle of each piece of your luggage and put at least your large, packed luggage outside your cabin door...usually by 10 or 11 PM. It will be taken by stewards for you to claim ashore in the terminal the next morning.
- Unlike airlines, you can put out more than two pieces per person. But if you're flying home, you'll need a way to deal with all pieces somewhere ashore before your flight.
- You can also put out carry-on baggage, but...
- Don't pack in them what you'll need in the morning, *e.g.*, medications, toiletries, something to wear to leave the ship. This often means you'll need to later transfer some items (especially liquids) you've kept for use into your to-be-checked luggage...sometime before you approach the point where you'll flight-check your bags.

Always Keep with You any Bags that have Valuables Inside.

Some ships offer a "walk off" or "express" disembarkation option. You can choose to leave the ship early as part of a group receiving special, expedited C and I processing if you carry off all your possessions. You should find that option on the questionaire mentioned above. If you opt for it, do not miss the call for your group to disembark or you may be directed to leave in the very last group.

At some major ports, some ships offer special luggage handling services for passengers flying home on certain airlines. For a fee, they will provide special airline tags and take luggage (you intend to check) the last night to, perhaps through your airline check-in at the airport. If you opt for this, understand the process and your responsibilities.

Typically, people with early flights, cruise-line shore excursions and express walk-offs are given the earliest debarkation times. The next/last morning, you may be instructed to leave your cabin somewhat early (yes, a dining room and the buffet will open quite early for breakfast), and proceed to a specific public area on the ship to wait to be called to disembark. (Per "Get in" above, your cabin steward must completely redo it and many others by 2 PM or so for the next occupants.)

- Consider having a hearty breakfast. You may face several hours before the next opportunity for decent food.
- If anyone in your party needs to be seated while waiting (perhaps 1-2 hours) to disembark, they should go your assigned public area early.
- That area won't necessarily be on the same deck as for exiting the ship. If not, you may have to negotiate stairs with your belongings because elevators will be very busy. If anyone has physical limitations, request help in-advance.
- In that area, you'll eventually be called by colour/number to disembark...usually not before the scheduled time.
- Many ships do not call groups using the public address system. They're made only by staff in your assigned public area.
- If you're not part of a special group that leaves early, and you miss your group's call, you can usually leave later without complications.

As you walk off the ship into the terminal, you'll be ushered to a large area with masses of luggage...considerably less if you disembark in one of the last groups. Your luggage will be arranged according to your disembarkation group. You'll need to pick-out your pieces and proceed to C and I or other processing if any. Porters will likely be available to help. After any processing, many options begin, *e.g.*,:

- If you have an end-of-cruise tour or pre-paid airport transfer arranged, you'll be ushered to a bus or other vehicle to transport you, with your luggage usually on-board.
- Other transport can take many forms, *e.g.*, self-arranged taxis, surries to paid parking lots or local hotels.

Once you leave the terminal, you basically have full responsibility to reach whatever destination or follow-on transport awaits.

2

The Cruise Industry

Over the last decades, the modern cruise industry has responded to extensive market and consumer research with the presentation of innovative naval design concepts, new ship lengths, ever more exotic destinations around the globe, and new on-board and on-shore activities and themes, developed to offer a vacation alternative that satisfies the expectatives of everyone.

THE CRUISE INDUSTRY DEVELOPMENT

The cruise industry, which modern version dates from the 1970s with the development of the North American industry, has experienced an increasing process of popularization, becoming a major part of the tourism sector, and reaching a level of enormous significance world-wide as an economic factor. The modern cruise industry is also one of the most outstanding examples of globalization, with an increasing number of ports of call and destinations around the globe, a multinational clientele and onboard personnel from every continent, and a level of detachment from communities and nations never seen before in history, with important economic, legal, environmental and social implications.

A crisis-resistant industry with a diversified offer of airlift options and modernized port structures that have open up cruising as a vacation alternative available for an increasing, more affluent customer base, offering a exciting, eventful, relaxing and definitely enjoyable experience for millions of passengers from across the world each year.

This dynamic sector is continuously expanding its offer of products and services, and developing new markets, with an average 8.5 per cent annual growth in the last 20 years, and nearly 90 million passengers since 1980, 60 per cent percent of whom have been generated in the past decade, in a tendency that shows no sign of slowing, with 13 and 13.5 million passengers in 2008 and 2009, compared with the 12,6 million in 2007, and that is expected to continue through the 21st century. In terms of capacity, the cruise industry has also experienced unprecedented development since the turn of the century. During the 1980s, around 40 new cruise ships were built and put in service, followed by other 80 vessels along the 1990s, and a 40 per cent increase between 2000 and 2005 that will be completed by and additional 25 per cent of new state-of-the-art units to replace ships that are expected to be withdrawn in the next years.

A multimillion investment into new, more innovative and ever-bigger vessels capable of carrying up more than 3,000 passengers, offering lower fares and shorter cruises to benefit from economy of scale and onboard activities such as multi-story shopping centers, restaurants, cafés and pubs, nightclubs, discos, casinos, art galleries and museums, theatres and cinemas, libraries, personal care areas and spas, gyms, swimming pools, tennis courts, ice skating rings, and a long etcetera of amenities to meet the changing vacation patterns of today's market and exceed the expectations of its customers with practically a cruise option for everyone.

A fleet compossed of several hundred large cruise ships carrying millions of passengers plies routes in all geographical areas in an expanding range of more than 500 destinations worldwide, with the Caribeean cruises as the favourite ones, followed by Mediterranean cruises and European itineraries that visit diverse popular ports and cities – Barcelona, Venice, Nice, Athens and the Greek Islands, Monte Carlo, Istambul, London, Amsterdam, the Scandinavial Fjords, Helsinki, San Petersburg, etc. –, including also the opportunity to enjoy places not included in the usual offer presented by other travel and tourism service providers, such as Artic and Antarctic regions.

Indeed, the cruise industry has increased in popularity all around the world, serving an heterogeneus clientele with well-differentiated expectations and preferences in the Asian, European and North American markets. This phenomenal growth has also created the need for more efficient managerial, organizational and planning structures to best the increasing competition and deal with the many changing factors in an also evolving market that generates over $15 billion every year – 79 per cent of which corresponds to the North American and British markets – and hundreds of thousands of direct and indirect jobs around the world, yielding an indirect multi-billion dollar annual benefit in diverse industrial sectors in all the world (nondurable and durable goods manufacturing, professional and technical services, travel services, financial services, airline and transportation, and wholesale trade).

THE CRUISE INDUSTRY CONSOLIDATION

In the other hand, the growth of the cruise industry has been severely conditioned by diverse events, such the Achille Lauro hijack in 1985, the Iraq and Kosovo wars, and especially the aftermath of the September 11 attacks, followed by an accentuated process of restructuration and consolidation in the sector. Renaissance Cruises was the first company to fill for bankrupcy on 25 September 2001, followed by American Classic Voyages and other four companies reflecting ten well-known brand names, ceasing operations and leaving the market wide open for the largest cruise companies – Carnival Corporation, Royal Caribbean Cruise Limited and Star Cruises – to consolidate in a process of purchasing and merging of companies giving birth to enormous corporations that control about 80 per cent of the cruise market worldwide.

Carnival Corporation, headquartered in Miami (Florida, USA) and London (England, UK), is the undisputed leader in the sector and the most profitable leisure company in the world, with 12 cruise brands in North America, Europe and Australia that operates 89 cruise ships, around 65,000 shipboard employees and 170,000 guests at any given time.

Among the companies of more modest size in the market are Crystal (subsidiary of the Japanese NYK), Silversea and Raddisson Seaven Seas. In the other hand, althought the corporate offices of the main cruise companies are located in the United States and Europe, and their clientele comes mostly from these same areas, these organizations, as an integrant part of the shipping industry, are incorporated and have most of their fleet registered in Panama, Liberia, Bemuda and Bahamas to obtain a series of benefits and advantages that allow a better economic balance and competitiveness derived from more favourable standards concerning taxation, labour laws and safety and environmental regulations.

THE CRUISE INDUSTRY COMPETITION

The volume of the cruise ship market is relatively small, with important barriers both to entry and exit associated with the extremely high cost of purchasing or selling a single cruise ship, and the high investment needed to maintain and manage a cruise line, which has a decisive influence on diverse aspects and strategies related to organizational and management issues. In the same way, the bargaining power and capability to take advantage of economies of scale present in the cruise industry are also affected by the size of the market in two different and contradictory ways: a) the presence of a few shipbuilders and techology developers in the industry forces the cruise companies to accept the prices and costs offered to them; b) the large number of suppliers of equipment, fuel and food products allows them to bargain for the best prices.

Such a reduced number of companies allows them to watch closely for potential commercial threats in a constant competition for a clientele influenced by general economic conditions and with other vacation alternatives, such sightseeing vacations, land-based resort hotels, thematic parks, etc. A situation that has given place to diverse strategies and plans to identify and specialize in the specific areas within this business framework, resulting in a division of the cruise industry into well-differentiated sectors or market niches – luxury, premium and contemporary – that offer diversified and targeted cruise products and services to satisfy both mass consumer markets, interested in budget packages, and a distinctive clientele seeking the exclusive environment onboard small ultra-luxury ships. Marketing, innovation and brand image are therefore vital elements in such a competitive commercial environment, and key factors to succeed in a sector in a constant effort to find new sources of income and new strategies to maximize economic performance and profit.

CRUISE INDUSTRY REVENUE

The main revenues in the cruise industry are generated for the most part from cruise ship passengers, and the ability to attract and maintain a clientele is therefore essential to its financial success. However, cruise fares are currently just another element in the complex mosaic of commercial relations around the cruise activity, quite far already from the relatively inexpensive or initial all-inclusive vacation packages offered in the 1970s, with a growing number of strategies for generating onboard revenue. The time when onboard shops provided a few items to complete passengers' luggage, some souvenirs and duty-free products was left behind long ago. Never to return. Nowadays, cruise ships offer an increasing range of shipboard stores and boutiques, spa and personal care services, photography departments and art auctions at prices that compete with land-based establishments. Most of them provided and managed by concessionaires and subcontractors.

Cruise companies have also introduced diverse practices to attract customers to spend their money, for instance, in extra-tariff and alternative restaurants and bars, satellite telephone services, cybercafés, and diverse revenue-generating schemes in passengers' cabins (interactive multimedia and TV, minibars, etc.). In addition, while many people go on a cruise with the intention of doing nothing more than relaxing and unwinding; other customers are interested in participating in all sorts of activities and experiencing the destinations to the fullest, and willing to pay for it. As a result, there is an increasing offer of shipboard activities and services, such as Bingo and casino gambling, adventure sports, culinary workshops, videogames, computers and virtual reality centres, theme nights, etc.

In the same way, cruise passengers are also encouraged to take part in a growing number and variety of auxiliary onshore activities. In the early 1990s, cruise companies began marketing a diversity of onshore activities and services. Since then, guided excursions and port lecturers, contracted with local concessionaires and local tour operators to be later sold to passengers onboard, have become the largest growing source of income for some cruise companies. Cruise-based tours of several hours while the ship is docked at a port of call or anchored a few miles offshore, which offer different themes and exciting activities: sightseeing, with excursions to natural, ecological and biosphere reserves, and protected areas that include wildlife viewing; adventure tours, adventure sports excursions and diverse activities in privileged natural environments; and historical and cultural based tours, with a strong educational component derived from the opportunity to visit museums and monumental heritages. The cruise ship industry has also shown an ability to establish and maintain effective relations with the land-based tourism industry. Moreover, the increasing purchasing and bargaining power of cruise companies has a significant impact on the providers of these services, competing and forced to undercut one another to successfully secure a contract with a cruise company, which allows the industry to obtain an additional income from the difference between purchasing and selling prices of such products and services.

On-shore excursions and visits to ports generally provide an extra income to cruise companies. Thus, cruise passengers usually receive a map to illustrate about the most recommended itineraries, including a buying guide that identifies a listing of stores and commercial establishments in the area, approved by the cruise company on the grounds of good prices and guarantee of quality. Not everyone knows that to be approved and included in the listings, those same stores and business have had to pay an upfront fee or agreed to share a certain percentage of the passengers' purchases.

In 1990, these strategies and policies were taken to a next level with the introduction of the concept of "private island", developed by Norwegian Cruise Lines and adopted by other companies operating mostly in Caribbean waters, with different economical benefits by providing an alternative to traditional congested ports, including a monopolistic control over local stores and services, open and available round-the-clock while the cruise ship is docked, which eliminates any competition and provides a controlled scenario that ensures a positive experience to their clientele, with the consequent additional benefit in the form of positive referrals to other potential customers.

Leaving aside onboard revenue, the cruise industry has other alternative means to improve their economic results: reduction in their costs by economies of scale and implementation of improved management systems (already mentioned above), and participation in the economic benefits of ports and port-related activities. Port charges have become an interesting source of income for the cruise companies since the 9/11 attacks, when many companies redesigned their itineraries and destinations closer to the United States, which opened a market for diverse Caribbean cities willing to offer a reduction in their port charges or a bounty for each passenger to attract cruise ships to their ports. In a similar way, there have been many initiatives from cruise companies to help economically in the building or improvement of port facilities and infrastructures in return for a future revenue-sharing formula, which would include, in some cases, priority berthing and a percentage of port charges.

PREDICTIONS

On a short-time basis, and in spite of the potential growth of the cruise industry and its capability to move ships and switch itineraries so as to adapt to the evolving demand, the rising of fuel prices, the current economic crisis and diverse armed conflicts and political instability that affect diverse parts of the world are taking their toll on this sector. For the last decades, cruise line companies have been ordering new and innovative ships yearly. Nevertheless, the new cruise ships coming within the next years, valued in more than 20 billion dollars to bring additional 85,480 berths into the market with estimated 4.2 million passengers by 2012, were ordered when the dollar had a more favourable exchange rate, and some companies are already cutting back their fleet and revising their order plans downward past that date.

In the other hand, other companies cruise companies, as Royal Caribbean International, have placed orders for the largest, more luxurious and innovative ships ever to be constructed, included in the Genesis-class vessels, with a cost of $1.65 billion, 220,000 GT and a capacity for 5,400 passengers and around 2,100 crewmembers, including a series of innovative shipboard amenities and facilities in the next years. In any case, the current economic situation has made the companies in the sector reconsider their business plans, controlling costs and reducing part their staff.

However, these economic adjustments and changes will not have significant effects on consumers and their access to high quality cruise services. Thus, the possible lack of new large cruise ships after 2012 is already been compensated by many companies in the framework of ambitious refurbishing programmes to keep their fleets equipped with innovative onboard offerings to deliver a memorable experience to their passengers; together with the design and construction of boutique-sized ships with a few hundred berths, equipped with a full set of luxurious amenities, competing to offer a more intimate environment. Moreover, many cruise executives are convinced that the current situation involves a certain beneficial effect on the industry, based on the actual cost difference with other alternative vacations. Thus, cruise vacations are more affordable than other land-based alternatives, including transport, accommodation, meals and entertainment in the final price, with more and more holidaymakers interested in buying a cruise vacation every year.

Following this trend, the cruise industry is looking for new initiatives to increase its clientele while maintaining the current market share, and each company has its own strategy for coping with new challenges, offering, for instance, multigenerational family travel, more innovation in the entertainment offerings, new onboard activities, more specialization in the offer of services, and aggressive pricing models and discounts to lure potential travellers.

In the same way, progressive concentration in the sector is expected to continue increasing, threatening the survival of regional companies not specialized in a concrete market niche, including also the predictable appearance and development of new large-scale companies competing for the emerging Asian markets; while smaller, newer markets in Europe and the Middle East,

Amazon and Brazil, Greenland and the Antarctic regions, also offer prospects for long-term growth. In the other hand, many economies and ports in all the world look at the cruise industry as a potential source of development and economic growth; and many companies have already announced plans to add new ports of call to existing and new itineraries, including coastal and river cruises, and many cruise options from more than 30 domestic ports just in the United States and Canada, providing a better road accessibility to the embarkation ports, and the consequent saving in airfare.

While not immune to the current economic situation, most cruise companies and travel agents are expressing optimism for the economic outlook of the sector when looking ahead over the next years. In any case, the industry has to deal with diverse problems and challenges that adversely affect the demand for cruise vacations alternative and, consequently, its operating margins and profitability. Therefore, environmental issues and concerns regarding living and working conditions onboard cruise ships, onboard safety, safety and health, and many other unresolved subjects claim to be taken in an ethically responsible way if the cruise industry is to maintain its quality of service in a framework of sustainability and competitiveness in the next decades.

A BRIEF HISTORY OF THE PASSENGER SHIP INDUSTRY

The earliest ocean-going vessels were not primarily concerned with passengers, but rather with the cargo that they could carry. Black Ball Line in New York, in 1818, was the first shipping company to offer regularly scheduled service from the United States to England and to be concerned with the comfort of their passengers. By the 1830s steamships were introduced and dominated the transatlantic market of passenger and mail transport. English companies dominated the market at this time, led by the British and North American Royal Mail Steam Packet (later the Cunard Line). On July 4, 1840, Britannia , the first ship under the Cunard name, left Liverpool with a cow on board to supply fresh milk to the passengers on the 14-day transatlantic crossing. The advent of pleasure cruises is linked to the year 1844, and a new industry began.

During the 1850s and 1860s there was a dramatic improvement in the quality of the voyage for passengers. Ships began to cater solely to passengers, rather than to cargo or mail contracts, and added luxuries like electric lights, more deck space, and entertainment. In 1867, Mark Twain was a passenger on the first cruise originating in America, documenting his adventures of the six month trip in the bookInnocents Abroad. The endorsement by the British Medical Journal of sea voyages for curative purposes in the 1880s further encouraged the public to take leisurely pleasure cruises as well as transatlantic travel. Ships also began to carry immigrants to the United States in "steerage" class. In steerage, passengers were responsible for providing their own food and slept in whatever space was available in the hold.

By the early 20th century the concept of the superliner was developed and Germany led the market in the development of these massive and ornate floating hotels. The design of these liners attempted to minimize the discomfort of ocean travel, masking the fact of being at sea and the extremes in weather as much as possible through elegant accomodations and planned activites. The Mauritania and theLusitania, both owned by the Cunard Line of England, started the tradition of dressing for dinner and advertised the romance of the voyage. Speed was still the deciding factor in the design of these ships. There was no space for large public rooms, and passengers were required to share the dining tables. The White Star Line, owned by American financier J.P. Morgan, introduced the most luxurious passenger ships ever seen in the Olympic (complete with swimming pool and tennis court) and Titanic. Space and passenger comfort now took precedence over speed in the design of these ships-resulting in larger, more stable liners. The sinking of the Titanic on its maiden voyage in 1912 devastated the White Star Line. In 1934, Cunard bought out White Star; the resulting company name, Cunard White Star, is seen in the advertisements in this project.

World War I interrupted the building of new cruise ships, and many older liners were used as troop transports. German superliners were given to both Great Britain and the United States as reparations at the end of the war. The years between 1920 and 1940 were considered the most glamorous years for transatlantic passenger ships. These ships catered to the rich and famous who were seen enjoying luxurious settings on numerous newsreels viewed by the general public. American tourists interested in visiting Europe replaced immigrant passengers. Advertisements promoted the fashion of ocean travel, featuring the elegant food and on-board activities. Cruise liners again were converted into troop carriers in World War II, and all transatlantic cruising ceased until after the war. European lines then reaped the benefits of transporting refugees to America and Canada, and business travelers and tourists to Europe. The lack of American ocean liners at this time, and thus the loss of profits, spurred the U.S. government to subsidize the building of cruise liners. In addition to the luxurious amenities, ships were designed according to specifications for possible conversion into troop carriers. Increasing air travel and the first non-stop flight to Europe in 1958, however, marked the ending of transatlantic business for ocean liners. Passenger ships were sold and lines went bankrupt from the lack of business.

The 1960s witnessed the beginnings of the modern cruise industry. Cruise ship companies concentrated on vacation trips in the Caribbean, and created a "fun ship" image which attracted many passengers who would have never had the opportunity to travel on the superliners of the 1930s and 1940s. Cruise ships concentrated on creating a casual environment and providing extensive on-board entertainment. There was a decrease in the role of ships for transporting people to a particular destination; rather, the emphasis was on the voyage itself. The new cruise line image was solidified with the popularity of the TV series "The Love Boat" which ran from 1977 until 1986.

A BRIEF HISTORY OF THE CRUISE LINE INDUSTRY

Before the dominance of air travel, which began to enjoy commercial success in the late 1960s, passenger liners were the preferred mode of overseas travel. Ships changed very little during the first half of this century. Although engine efficiency improved, passenger staterooms, public lounges, and deck space on a cruise ship built in the 1950s were not much different than those on the S.S. Titanic. Most of the so-called modern ships plying the waters during this time copied the amenities and grand styling of past steamships. Their purpose remained the same as well. Oceangoing vessels were primarily used to get from Point A to Point B, especially for second- and third-class passengers, whose accommodations were in stark contrast to those in first-class staterooms. The most common voyages were transatlantic crossings from New York to London.

Interestingly enough, you can still visit the original terminal at Tilbury Docks, just east of London. This is a fun way to see the beginning of the cruise ship industry while imagining the hustle and bustle of a bygone era. It sits humbly on the tide flats of the Thames, surrounded by the smoke stacks of the city. It was from here that the grand Queen Elizabeth ran continuous service to New York for so many years. In this golden age of ocean liners, ships such as the Lusitania, United States, Ocean Monarch, Paris, Queen Mary, Caronia, and Laconia were sailing the high seas. Every vessel had a unique personality, history, and enough passengers to keep companies financially afloat. But it was an age that would eventually come to an end.

The real blow to the cruise ship industry came in the 1960s when Boeing began selling 747s and other aircraft worldwide. Meanwhile, a global transportation network of airports with regulated common language and air traffic controllers, in coordination with the United States Federal Aviation Administration, was being established. As the decade came to a close, it was no longer fashionable, practical, or economical to travel by boat. The age of the jumbo jet had arrived.

Who is to say which factors brought the concept of the cruise ship back to life. It is speculated that shows such in the 1970s classic The Love Boat contributed to the idea of luxury cruise ship travel and that ideas found in this show contributed to a growing trend of cruise ship travel as a luxury vacation rather than a means to travel as a necessity. With Princess Cruises focused on Caribbean itineraries, the popularity of cruise ship vacations grew by leaps and bounds. This also marked a time when cruise ship amenities began to change. Cruise ship travelers of all social classes would enjoy first-rate rooms and service. Ocean views, pools, casinos and onboard entertainment are all staples when it comes to cruise ship travel, as was the case in the late 70s when more and more cruise ships were being built. The same is true today. Larger and larger vessels are being constructed – a trend that has continued in the cruise ship industry as passengers see bigger and better ships.

EMERGENCE OF THE CRUISE INDUSTRY

From the mid-19th century liner services supported long distance passenger transportation between continents, particularly between Europe and North America. The need to accommodate a large number of passengers of different socioeconomic status for at least a week led to the emergence of specific ship designs radically different from cargo ships where speed and comfort (at least for the elite) were paramount. The emergence of the cruise industry can be traced to the demise of the ocean liner in the 1960s as it was replaced by fast jet services for which it could not compete. The last liners became the first cruise ships as it took more than a decade to see the complete demise of liner services with the final realization that long distance travel was now to be assumed by air transport and also considering the 30 years lifespan of a liner. The availability of a fleet of liners which utility was no longer commercially justifiable incited their reconversion to form the first fleet of cruise ships.

For instance, one of the last purposely designed liners, the SS France, operating between 1961 and 1974, was mainly used for the conventional transatlantic service between Le Havre and New York. With rising oil prices and more efficient jet liners, including the 747 (introduced in 1970), the liner was no longer able to effectively compete over the transatlantic route. While a jet plane could link Paris or London to New York in about 8 hours, it took about 4 days for a liner to cross the Atlantic, excluding a train segment between London and Southampton (or Paris and Le Havre). Considering one round trip per day, a 747 could carry about 3,200 passengers across the Atlantic in the time it took the SS France to carry 2,000 passengers on a single journey. Unable to generate enough revenue to justify its operating costs the SS France was mothballed in 1974 and purchased by Norwegian Cruise Line (renamed the SS Norway). Its final commercial years between 1980 and 2003 were spent as a cruise ship. However, liners were not particularly suitable to the requirements of the emerging cruise industry. For instance, since many liners were designed to operate on the North Atlantic throughout the year for scheduled passenger services, their outdoor amenities such as boardwalks and swimming pools were limited. Additionally, they were built for speed (which was their trademark) with the related high levels of fuel consumption.

The modern cruise industry began in the late 1960s and early 1970s with the founding of Norwegian Cruise Line (1966), Royal Caribbean International (1968) and Carnival Cruise Lines (1972), which have remained the largest cruise lines. The early goal of the cruise industry was to develop a mass market since cruising was until then an activity for the elite. A way to achieve this was through economies of scale as larger ships are able to accommodate more customers as well as creating additional opportunities for onboard sources of revenue. The first dedicated cruise ships began to appear in the 1970s and could carry about 1,000 passengers. By the 1980s, economies of scale were further expanded with cruise ships that could carry more than 2,000 passengers. The current large cruise ships have a capacity of about 6,000 passengers. The market for the cruise industry was by then established and recognized as a full-fledged touristic alternative directly competing with well-known resorts areas such as Las Vegas or Orlando.

Cruise ships tend to have a low draft since they do not carry cargo; they are more volume than weight. This confers the advantage of being able to access a large number of ports and therefore multiplying itinerary options since the setting of a pure cruise port leans on criteria that are different from commercial ports. Cruise ports tend to be located close to either city centers (cultural and commercial amenities) or to natural amenities (*e.g.* a protected beach). These sites do not have on average very deep drafts and dredging would be socially or environmentally unacceptable. For instance, ships of the Oasis class, which as of 2011 accounted for the largest cruise ship class, have a draft of 31 feet. Comparatively, a containership of 2,500 TEUs requires a draft of 33 feet, while a sovereign class containership of 8,400 TEUs requires a draft of 46 feet. Draft issues that have plagued container ports are a much more marginal constraint for cruise shipping. Additionally, cruise ships have the option to anchor and use tendering services, which opens a wide array of ports of call.

MARKET DYNAMICS

The global cruise industry offers a combination of local destinations accessible through the cruise port and on-board amenities offered by the cruise ship. It carried about 20.1 million passengers in 2012, up from 7.2 million in 2000. The global growth rate of the cruise industry has been enduring and stable, at around 7 per cent per year in spite of economic cycles of growth and recession. For instance, the financial crisis of 2008-2009 has not impacted the demand for cruises.

Three major trends have particularly shaped the cruise market since its emergence:

- Amenities. The cruise ship has become an important part of the cruise experience since it also represents a destination in itself. An increasing amount of amenities are being offered both on the ships and as shore experiences.
- Massification. The principle of economies of scale and the development of a wider customer base has incited the deployment of

larger cruise ships. While in the 1990s, cruise ships rarely exceeded 2,000 passengers, by the 2010 ships of 6,000 passengers were being deployed. Also, larger ships are able to support a wider range of amenities.

- Concentration. The industry has a high level of ownership and market concentration. Carnival and Royal Caribbean are the two main cruise lines and account for 73 per cent of the market. The Caribbean and the Mediterranean account for about 70 per cent of the global annual deployment of cruise capacity.

These trends underline that the industry has been so far fundamentally supply based; the ships are built and the customers are found to fill them through various marketing and discounting strategies. The possibility for cruise ship operators to successfully follow a supply push strategy makes the cruise industry quite different from other shipping markets, such as container shipping. This is also in contradiction to the tourism sector in general which is highly demand derived and thus sensitive to the general economic context. Hence, in most shipping markets the shipping activity is a clear derived activity of trade and demand is rather price inelastic. As mentioned earlier, demand in the cruise business is created through pricing and branding/marketing. Cruise operators are challenged to develop competitive cruise packages which involve a high-quality stay onboard, an array of shore-based activities offering access to a variety of cultures and sites and easy transfers to/from the vessel. Most cruise lines have a logistics office that is responsible to supply ships, particularly in terms of food, a core amenity. Variation in preferences are observed depending on the itinerary, the composition of passengers, including factors such as origin and age.

The construction of cruise ships tends to take place in cycles where several ships are ordered and enter the market within a short time frame. Since the cruise industry is a relatively small segment of the touristic sector, it has so far been very successful at finding customers to fill a greater number of ever larger ships. The cruise product has become diversified to attract new customers and to respond to the preferences of a wide array of customer groups. In view of fulfilling the desires of its guests, the cruise industry has innovated through the development of new destinations, new ship designs, new and diverse onboard amenities, facilities and services, plus wide-ranging shore side activities. Most cruise ship operators work around specific cruise themes and voyage lengths can vary to meet the changing vacation patterns of customers. Its highest level of market penetration is in North America with about 3 per cent of the population taking a cruise each year. This includes people who may take more than one cruise in a year so actual figures are actually lower.

The market drivers of the cruise industry are similar to those that have fostered the growth of tourism after World War II, particularly the rising affluence of the global population and the growing popularity of exotic and resort destinations. The general aging of the population is also a factor in favour of cruise shipping as the main market remains older adults, albeit customers are getting significantly younger. While in 1995 the average age of a cruiser was about 65 years, this figure dropped to 45 years by 2006. What is novel with cruising is that the ship represents in itself the destination, essentially acting as a floating hotel (or a theme park) with all the related facilities (bars, restaurants, theaters, casinos, swimming pools, etc.). This permitted cruise lines to develop a captive market within their ships as well as for shore-based activities (*e.g.* excursions or facilities entirely owned by subdiaries of the cruise line).

Some cruise operators go very far in developing new entertainment concepts on board of their vessels, including surf pools, planetariums, on-deck LED movie screens, golf simulators, water parks, demonstration kitchens, multi-room villas with private pools and in-suite Jacuzzis, ice-skating rinks, rock-climbing walls, bungee trampolines and other. Onboard services account between 20 and 30 per cent of the total cruise line revenues. The average customer spends about $1,700 for their cruise, including ship and off-ship expenses for goods and services. The majority of these expenses are captured within the cruise ship as passengers spend on average $100 per port of call. The Caribbean has been the dominant deployment market of the cruise industry since its inception, but the Mediterranean cruise market has grown substantially in recent years. Both markets offer a variety of cultures in close proximity and are thus ideality suited. Furthermore, strong niche markets have developed focusing on, for instance, history (Hanseatic cities in northern Europe) or natural amenities (Alaska).

The cruise industry has a very high level of ownership concentration, since the four largest cruise shipping companies account for 96 per cent of the market (Carnival Lines, Royal Caribbean, Norwegian Cruise Line and MSC Cruises). High levels of horizontal integration are also observed since most cruise companies have acquired parent companies but kept their individual names for the purpose of product differentiation. For instance, Royal Caribbean Cruises, which is the world's second largest cruise company behind Carnival Lines, accounts for 24 per cent of the global market serviced under 6 different brands such as Celebrity Cruises (which caters to a higher end customers) and Azamara Club Cruises (smaller ships servicing more exotic destinations with shore stay options). Like its cargo shipping counterparts the cruise industry relies on flag of convenience to gain fiscal advantages but mostly to secure more lax labor regulations since it is an industry relying extensively on labor. For instance, Carnival Cruises and Princess Cruises have all the ships registered in foreign countries (Panama, Bahamas and Bermuda).Cruise shipping is increasingly capital intensive as each new cruise ship class comes with better amenities. A ship of the latest Oasis class, which is able to carry more than 6,000 passengers, costs about 1.2 billion dollars and can take 4 years to be delivered. Larger ships command higher booking prices since they offer more amenities, but current trends indicate that the cruise industry has no ships larger than the Oasis class in its order books. Optimal economies of scale may have been reached.

NETWORKS AND PORTS OF CALL

The cruise industry sells itineraries, not destinations, implying a greater flexibility in the selection of ports of call. Over this issue, cruise ports come into two major categories; turn ports (or hub ports) where passengers begin and end their cruises and ports of call that are visited along an itinerary. The selection of an itinerary is the outcome of several commercial considerations including potential revenue generation, distance between ports of call (cruise ships can cover 200 nautical miles per night), brand positioning (exotic ports of call for premium services), guest satisfaction (customer oriented industry), economic trends and market research such as evaluating changes in disposable incomes and the demographics of the customer base. Ships are constantly moving between ports of call and shore leaves are of low duration; 4.3 hours on average in the Caribbean. A standard cruise itinerary is a loop beginning and ending at a hub port (also called a turn port or port of embarkation) and typically lasting 7 days with 3 to 5 ports of call depending on their respective proximity. Cruises of 10 to 21 days are also offered but they tend to have lower profit margins as customers are inclined to spend less as the cruise progresses.

Three main types of itineraries can be found:

- Perennial. The region covered by the itinerary is serviced throughout the year as the demand remains resilient, which is associated with stable (subtropical) weather conditions as well as stable itineraries. There may be significant seasonal variations in the number of ships deployed but the market remains serviced throughout the year. The Caribbean is the foremost perennial cruise market (summer low season), but the Mediterranean is also serviced year-round with a winter low season.
- Seasonal. Weather is the dominant factor explaining seasonality, implying that some regions have a market potential only during a specific period or season. This is particularly the case for Baltic, Norwegian, Alaskan and New England cruises that are serviced during summer months. Inversely, South American and Australian itineraries are serviced during the winter months.
- Repositioning. Because of the seasonality of the cruise industry the repositioning of ships between seasons is required. Cruise companies are increasingly using this opportunity to offer customers lower costs cruises for the inconvenience of having to book air travel

arrangements for the return trip since the beginning and ending ports of call are not the same. This mainly takes place across the Atlantic as ships move from the winter Caribbean peak season to the summer Mediterranean peak season (and vice-versa). The beginning and the end of the Alaska season are also combined with a Hawaiian cruise as ships get repositioned. Barcelona and Dubai are emerging repositioning hubs since the Mediterranean and the Indian Ocean are growing faster than the conventional Caribbean market.

For most customers, a cruise involves two travel segments, the first being air travel to the hub port (with a return trip) and the second is the cruise itself. It is therefore important that the hub port is serviced by a well-connected airport, with significant airlift capacity and which represents in itself a touristic destination. This is the case for Miami, Fort Lauderdale and San Juan that are respectively well connected airports and act as hub ports for Caribbean itineraries. Barcelona and Civitavecchia (near Rome) are major hub ports for the Mediterranean which are well serviced by air transportation. Poorly connected airports are commonly associated with higher airfares, which impair the competitiveness of the city for mass tourism. There are a number of customer benefits linked to having more cruise embarkation points available such as drive-to convenience (particularly in felt in North America) and fewer airport hassles. More "close to home" ports (also called "drive to ports") also increase the likelihood of cruising, the reason why cruise line will call ports along the American Gulf Coast and Eastern Seaboard such as New York, Tampa, Galveston, Baltimore and New Orleans.Environmental considerations play a role, particularly when calling at ports.

As a cruise ship carries several thousand passengers and crew in an enclosed facility, an array of municipal utilities such as water supply, electric generation, sewers and waste management are found on board. Large differences in CO_2 emissions can be observed between individual cruise ships depending on the size, the age and the ship's capacity configuration (*i.e.* high end luxury cruise ships vs. "mass" cruise ships). The largest ships show the lowest CO_2 output partly because of the high occupancy rate in number of beds per surface unit and their relative young age. Cold ironing or shoreside power facilities are being installed in a number of urban cruise terminals in view of reducing the environmental impact of docked ships. In 2001, the port of Juneau in Alaska was the first in the world to offer shoreside power for cruise vessels.

Seattle followed with two installations in 2005 and 2006. In 2009, Port Metro Vancouver also introduced a system to connect ships to the power grid so they can turn off their engines while docked. It is estimated that for an average cruise ship some 17,000 litres of fuel can be saved in a 10 hour docking period. In October 2010, San Francisco became the first port in California to offer clean shore power for cruise ships. Los Angeles, San Diego and Long Beach now offer similar facilities. Also in Europe, a number of ports have taken initiatives in this area (*e.g.* Gothenburg, Venice, Barcelona, La Spezia, Civitavecchia, and Hamburg).

The global cruise port system shows a high level of concentration and clustering. Cruise ports come into three main categories depending of the role they serve within their regions. They can be a destination port where there are few, if any, excursions taking place outside the port area, agateway port that mostly serves as a point of embarkation or a balanced port offering a combination of port area amenities as well as inland excursions. Each of these categories implies different development strategies to service the market. There has been a growing number of hub ports where passengers in whole and in part can begin or end their journey, so the future of the cruise industry main include more itineraries that are partially undertook by passengers. The cruise industry is expanding to provide more options for passengers, particularly for niche markets where higher prices can be commended than in the competitive mass markets of the Caribbean and the Mediterranean. For instance, cruises are set up for the Antarctic, the Hudson Bay and the South Pacific. In some cases, destinations may have more impacts than itineraries, inciting cruise lines to offer longer shore stay options where the cruise ship can become a temporary hotel in proximity to central areas. This is particularly the case when a major event is taking place such as the Rio Carnival, the Monaco Grand Prix or the British Open.

EMERGING CAPACITY CONSTRAINTS?

The siting and setting of cruise port facilities is subject to constraints usually not found for regular port activities. Many cruise ports are reconversion of port facilities that were used for cargo operations, particularly if those piers are adjacent to downtown. The seasonality of many cruise itineraries implies that an array of cruise ports are used during a peak season, but may receive limited, if any, cruise ship calls outside the main season. Cruise terminals can thus represent an underutilized asset prone to risks. For many ports, the cruise terminal is thus a temporary facility with the berth(s) supporting other uses when cruise ships are not calling. An approach is to use a cruise terminal as a multipurpose facility, often built for another rationale (such as a world event), that can include an adjacent hotel, convention center or any other revenue generating infrastructure that would gain on being on the waterfront.

An emerging trend, where possible, has been the setting of dedicated facilities where the cruise shipping company is directly involved in the development of the cruise terminal as well as co-located amenities. In some instances, private ports of call reserved exclusively for a cruise company have been developed, such as Labadee (Royal Caribbean) in Haiti, Coco Cay (Royal Caribbean), Half Moon Cay (Holland), Castaway Cay (Disney), Princess Cay

(Princess) and Great Stirrup Cay (Norwegian) in the Bahamas. These private facilities are all within one cruise day from the home ports of Florida, offering the option of short 3-4 days cruises to a quiet and safe destination. They are all enclaves without their respective host countries. The cruise industry has emerged to become a significant niche to the global tourism industry. The selection of ports of call and itineraries are carefully pondered to maximize the commercial potential and utilization of the ship assets.

From a market perspective, the cruise industry has the following unique characteristics usually not found in other segments of the tourism industry:

- Supply push strategy of cruise operators as they aim at "creating" demand simply by providing new capacity (ships) and finding customers to fill them.
- Offer itineraries where the whole is essentially greater than the sums of its parts. Specific regional and cultural experiences can be offered through a combination of sailing time and choice of ports of call.
- Expand and capture revenue streams by offering on board goods and services as well as shore-based excursions.
- Adapt to seasonal and fundamental changes in the demand by repositioning their ships (seasonal) and changing the configuration of their port calls (fundamental). One significant change concerns the rise in fuel prices, which account for 15 per cent of the operational expenses. For instance, Cunard announced in 2012 that it was lengthening the crossing time of its transatlantic cruises by one day, from seven to eight days.

Since the cruise industry appears fundamentally to be driven by supply, it is likely that supply saturation, as opposed to demand saturation, will constrain future developments and impose a maturity on an industry that has continued to grow rapidly. While large hub ports have the capacity to accommodate additional port calls, it is the smaller "exotic" or "must see" ports that cruisers are seeking to visit that present challenges for additional capacity. Berth availability and the capacity of small communities to accommodate large tourist influxes of short duration has become a salient issue. However, the massification of the cruise ships is placing a scale and scope challenge for several cruise ports that are facing the dilemma of the generation of tourism income and the environmental and social disruptions large cruise ships are associated with.

This could lead to further market segmentation between large ports called by mega ships and smaller ports called by smaller ships offering a specialized cruise experience.This is likely to incite the additional involvement of the cruise industry in terminal operations, a trend that has already taken place with the setting of private port / resort areas. The next step will involve the development of new cruise terminals co-located with service amenities such hotels, attractions, condominiums and shopping malls. Paradoxically, a similar trend was observed in container shipping in recent decades as several shipping lines became, through parent companies, terminal operators. While a further fragmentation of itineraries is likely to take place, a closer integration between the cruise port and the cruise shipping industry is to be expected.

ARRIVING BY CRUISE SHIP-TIPS ON HAVING FUN

Each year, hundreds of thousands of tourists arrive at our modern Port for a brief visit to Cartagena. Cartagena de Indias was the most important city on the main sea route, then known as: "The Spanish Main". The city held vaults full of gold, silver, emeralds and other treasures looted from the indigenous peoples of the continent, and Central America. Annually, when good sailing months permitted, the hoards were shipped to the treasury of the King of Spain. Because of her stored wealth, Cartagena was high on the "hit list" for the pirates.

Entering the Harbor, is in itself, an interesting part of your visit, and traces the path of the early galleons. Your first indication that you are approaching Cartagena occurs miles from the city, as your ship majestically glides right between two ancient Spanish Forts near Bocachica (Small Mouth), and then mid bay between the Forts of: Castillo Grande and Manzanillo. As pirate ships appeared on the horizon, a under water chain was raised from the sea bed between the forts, tightened and held fast, with a large capstan. From this advantage the ship was pummeled with cannon shot from the two opposing forts. As you continue your approach to the city you will see additional fortifications including the mighty Fort San Felipe de Barajas, which arguably is the largest and most impressive fortification in the new world. The Port of Cartagena is one of the most modern in South America and one of the busiest in Colombia. The Port cooperates fully with Interpol (of London, England) and the DEA (Drug Enforcement Agency) of the United States. There are video cameras everywhere, as well as a large contingent of well armed and trained police, plus drug sniffing dogs. Please do not be alarmed. The "troops" are friendly to tourists, and they will help you in any way that they can. A primary reason for this heavy security results from the insatiable demand for drugs consumed in the United States and other prosperous countries of the world.

The first law of Economics is: supply will meet demand. First, there is demand. The U.S. problems relating to drugs is shared by the people of Colombia, where more than 40,000 have been killed in the past 35 years. Within this same period, a much higher number of drug related deaths, have occurred within the United States.

Both countries share a "bad name" when it comes to drug violence and killings. Your visit to Cartagena is very much appreciated, and a big help to the local economy. The vast majority of locals truly love and appreciate gringos, and others foreigners who visit their city. There are more than 80,000 Colombians residing within the United States. It is a sad note, but it can be safely said: the majority of youth in Colombia, would love to: "live and work in the states". For many, it is their daily dream!

Some of your shipmates will not get off the ship in Cartagena. Others, who are equally un informed, will venture off the ship, but remain in the port, so they can brag to their friends: "I was in Colombia". The usual response will be: "you were where?...are you crazy?" Cartagena is not representative of Colombia, and there is little to worry about when it comes to your personal safety. The city, and in particular, the old walled portion called El Centro, is clean, safe, fun and romantic. Those who are not as informed as you, will be the big losers if they miss this golden opportunity to visit one of the greatest cities in South America.

While on your Cartagena adventure, you will see poverty-up close. The U.S. State Department for years, has issued a: Travel Warning to U.S. Citizens against visiting Colombia. If you seek a more balanced and accurate report, visit the Canada Advisory. Canada has a consul in Cartagena. They know first hand about personal safety.

The Travel Warning issued by the U. S. State Department is seriously hurting the economy and people of Cartagena! If you are a U.S. citizen, you should feel good knowing that your visit will help our economy. You are welcome, and the people of Cartagena thank you.

Note: This text is written by a 62 year old U.S. citizen who lives and works in Cartagena. "Even with a so-so command of the Spanish language, I don't feel that I have ever been in personal danger in Cartagena. My work keeps me out at all hours, visiting clients, many times while I am alone. Of course, I am cautious as to where I go, the way I dress and act, but this is also true when I visit my home state of Florida." Use a little common sense, and you will have no problems. Cartagena, is the most popular tourist city in Colombia. Our poverty is demonstrated by the high number of street hustlers. Foreigners stand out. You can be spotted a block away. When you are in a crowd, such as on a shore tour, the word passes quickly, and your group will be besieged like bees heading for the flowers. Hustlers are poor, and tourists represent the rich honey pot-their next meal.

THE FUTURE OF THE CRUISE INDUSTRY

Fig. Designing Cruise Ships Today to Meet the Wants, Needs, and Desires of Future Generations.

Having just returned from a holiday cruise on the Mexican Riviera, I became intrigued with the prospects for this behemoth industry, and the long-term implications of designing ships today that mesh well with the changing attitudes and fickle interests of traveling consumers many years in the future. So for the past week I have immersed in researching the cruise industry and thinking through the current trends that will guide it into the future. At the heart of this discussion is the 30-50 year economic life of a ship. With design and construction time for mega-ships taking five to ten years, it is already possible for them to be unveiled to the public noticeably out of date.

And 5-10 years into their operational life, cruise lines run the risk of having to schedule a major rework to re-sync the functionality to match consumer expectations. Some interesting parallels can be seen with shopping malls, where mall owners have seen stiff competition from Wal-Mart, Target, and Costco and are now filing a record number of bankruptcies, and jumbo jets with a similar usable life span but a rapidly increasing cost of operation during the later years. Unlike shopping malls, cruise ships can easily be moved to better markets. And unlike jumbo jets, many repairs and upgrades can happen "on the fly." For the moment, the industry is sitting on top of the world with a growing consumer base and stellar earnings reports. It is precisely this time when the industry needs to spend considerable time reinventing itself. With this in mind, we begin this examination of future trends in the cruise industry.

INDUSTRY OVERVIEW

The cruise industry is the fastest growing segment of the travel industry – achieving more than 2,100 percent growth since 1970, when an estimated 500,000 people took a cruise.

- Roughly 14.3 million passengers traveled in 2010, a 6.3 per cent increase over 2009.
- The cruise industry is the fastest-growing category in the leisure travel market. Since 1980, the industry has experienced an average annual passenger growth rate of 7.4 per cent per annum.
- Since 1990, over 154 million passengers have taken a 2+ day cruise. Of this number, over 68 per cent of the total passengers have been generated in the past 10 years and nearly 40 per cent in the past 5 years.
- The average length of cruises is 7.2 days.
- The cruise product is diversified. Throughout its history the industry has responded to the vacation desires of its guests and embraced innovation to develop new destinations, new ship designs, new and diverse onboard amenities, facilities and services, plus wide-ranging shore side activities. Cruise lines have also offered their guests new cruise themes and voyage lengths to meet the changing vacation patterns of today's travelers.
- The cruise industry now has over 30 North American embarkation ports placing cruise ships within driving distance of 75 per cent of North American vacationers. With the added convenience of avoiding air travel, cruise lines have attracted a wider customer base.
- From a capacity standpoint, utilization is consistently over 100 per cent (104 per cent in 2009).
- The Caribbean is the number one destination, with an estimated 37 per cent of the total in 2010.
- 26 new state-of-the-art new ships are contracted or planned to be added to the North American fleet through 2012, at a cost of nearly $15 billion US.
- Because only approximately 20 percent of U.S. adults, and far less of the world market, have ever taken a cruise vacation, there remains an enormous untapped market.
- Ninety percent of all cruise vacations are booked through travel agents.
- There are over 2,000 ports of call around our planet that cruise ships can visit.
- There are more than 300 cruise ships in the world today with a collective capacity to handle over 250,000 passengers.
- Three major cruise line groups (Carnival, Royal Caribbean and Star/NCL) control roughly two-thirds of world's cruise passenger capacity.

PROFITABILITY

Ever wonder where the money goes? Based on market calculations provided by Cruise Market Watch, here is a breakdown of the estimated 2010 average cruise revenue and expense per passenger for all cruise lines worldwide.

INCOME (Per Passenger)

Ticket$1,155

On board spending

-
  - Casino and Bar206
  - Shore excursions (cruise line portion)75
  - Spa37
  - All other on board spending 56
  - Total on board spending375

Total Revenue$1,530

EXPENSES (Per Passenger)

- Corporate operating costs$ 525
- Shipboard payroll169
- Agent commission162
- Depreciation and amortization153
- Ship fuel costs107
- Victualing (subsistence supplies)96
- Interest expense76
- Other ship expenses (port fees etc.,)69
- Other on board operating costs67

Total Expenses$ 1,424

Profit before taxes$ 157

For 2011, the average profit per passenger per day is projected to increase to $218.57, with $165.00 ticket price and $53.57 on board spending.

SIGNS OF TROUBLE

To be sure, the industry is doing many things right, so to some of you, the problems I'm mentioning here may seem like minor blips on the radar screen of success. Indeed, the total worldwide cruise market is estimated at $29.4 billion, a full 9.5 per cent increase between 2009 and 2010.

Yet for companies that pride themselves on offering luxury class service and amenities, there are some glaring omissions:

- Internet Connections – In virtually every other aspect of the travel industry, vacations have transitioned into working vacations, at least on some level. But with excessive connection fees and slow download speeds, doing work on-board a ship is painful at best. The average person in the U.S. spends 2 hrs and 35 minutes per day online, and as cruise lines attempt to recruit a larger share of the traveling public, they will be butting up against some natural barriers here until they are able to solve this issue.

- Cellphone Connections – While the rest of the world has shifted from place-to-place communications to person-to-person communications, the cruise industry remains woefully behind. Cell phones and other handheld devices are not usable without paying exorbitant connection fees. This also means that friends and family members on-board become increasingly difficult to coordinate plans with. Especially on the megaships of the future.
- Ship Time – With the proliferation of self-correcting watches, clocks, and timepieces, the notion of running a ship's operation on "ship time" becomes more confusing. Most ships have very few clocks and passengers are left in a constant state of confusion as to whether their watch or handheld device has self-corrected to a different time zone. This becomes a critical issue when a passenger is off on an excursion and arrives too late for the departure.
- Smoking – On-board smoking issues will continue to plague the cruise industry for years to come as the anti-smoking zealots in the U.S. square off against the chain-smoking Asians who are beginning to wield far more clout as their market share increases. Smoking bars and smoking casinos become difficult environments to contain and since casinos are a high-revenue asset, and smoking gamblers spend far more than non-smoking gamblers, many cruise lines have positioned their casinos along critical must-walk-thru traffic lanes to maximize impulse gambling, but at the same time, force non-smokers to breathe the air.
- Business Environment Vs Vacation Environment - Cruise ships are a natural environment for networking, collaboration, and launching new ideas. Yet the current technological constraints provide a massive impediment for attracting the serious tech-related businesses crowds of the future. As an example, full-immersion events like Seedcamp, Startup Weekend, or BarCamp would work well with a ship's self-contained live-together, work-together environment, but not without access to high bandwidth Internet connections.

FUTURE TRENDS, FUTURE OPPORTUNITIES

For cruise lines it was important to first establish a durable industry, and they have done a remarkable job so far. But here is where it gets interesting. Each of the industry leaders are now well-positioned to leave their mark on the future, and they will be doing it by pushing the envelope, taking risks, and breaking rules. Cruise lines, while still lagging on the digital frontier, will be entering an experimentation phase with each trying to establish themselves as a leading innovator. Here are eight key trends that will begin driving this industry into the future:

Global Load Shifting

Over the years the cruise industry has grown up with a majority of its passengers coming from North America. However, with the fluidity of global markets causing a constant shifting in the wealth of nations, the North American dominance of the industry is beginning to erode. Cruise Lines International Association's (CLIA) latest report shows that North American passenger numbers rose 1.0 per cent to 10.1 million, while the global passenger count grew 3.3 per cent to reach 13.4 million.

Leading travel expert, Arthur Frommer recently cited the shifting of capacity to European waters "the biggest development in cruising" noting "you'll see far fewer cabins and berths in the Caribbean." At the same time, countries around the world who are interested in building their image are realizing that cruising is an important vehicle for sampling destination areas to which they may later return. Over the coming years, the rapidly growing Asian market will cause a constant realignment of ships and cruise strategies with companies moving more of their ships into European and Asian ports. Already positioned as some of the most preferred destinations for travelers, look for strong growth in the Australian, New Zealand, and Singapore cruise industry.

Branded Differentiation

When scanning through the current listings of cruise options, it's easy to become overwhelmed by the uniformity. They all start looking the same. Yes, they originate from different ports and make a variety of stops at diverse ports along the way, but there is little to set them apart. People today are more inclined to identify with a branded experience as opposed to a branded ship. With this trend we will begin to see cruise companies align themselves with corporate sponsors in an effort to make every 5, 7, or 14 day cruise a branded experience.

Here are a few examples:

- ELLE Fashion Week Cruise – Where top fashion designers and top fashion models from around the world meet to unveil their latest creations at this once a year event sponsored by ELLE Magazine
- The Petron Tequila Cruise – Tequila lovers, here's the perfect getaway. Not only are all tequila dinks onboard half-price, and all passenger gifted with a one-of-a-kind free gift package from Petron, but people will have the opportunity to take an extended behind-the-scenes tour of the Petron factory in Jalisco, Mexico, and meet some of the workers.
- The Letterman Cruise – Join David Letterman and his guests as the entire Letterman teams is assembled onboard to produce his nightly show. For this week, the ship has been temporarily renamed, the U.S.S. Letterman.

- X-Box Tournament Cruise – If you are into gaming you will love this cruise with six X-Box Tournaments happening simultaneously on virtually every level of the ship.
- Facebook Social Networking Cruise – Invite your friends to join you on this cruise and for every friend that signs up, you win prizes and credits good for future cruises. The top ten networkers will have their trips totaled paid for. Passengers will be joined by key executives at Facebook as they step you through what's next for this social networking giant.
- The Amazon Shopping Cruise – No, this isn't a trip into the Amazon. Rather, passengers will be treated to an ongoing series of product demonstrations of everything lining the online shelves on Amazon.com. As a major incentive, all passengers will receive a free Kindle Book Reader.
- The Zappos Shoe Cruise – People who take extended cruises know the value of having great shoes. Daily shoe fashion shows, lectures by shoe designers and fashion trend experts, and meet the Zappos team as they demonstrate how their company is being positioned to meet the footwear needs of future generations.
- NYSE Cruise – On this cruise, 25 traders from the New York Stock Exchange will be working live from a mini-trading floor on-board the ship with live data feeds providing real-time updates on stock prices.

Growing Need for Office Staterooms

As the pace of business continues to climb, and the nature of employment continues to morph, few will be able to completely escape the demands of work for the duration of a cruise. For this reason, companies will begin to redesign their staterooms to include a functional work environment.

As an example, ceiling mounted flat screen televisions will enable one person to watch TV with headphones and not distract someone else who is working on a computer at a desk. Rolling office chairs, pull-out desk space, affordable in-room Internet and cellphone connections, projection screens, and large-screen computer monitors are just a few of the elements needed to give travelers the convenience of on-demand workspace whenever the need arises.

Rapidly Evolving Shipboard Innovations

Smaller ships will tend to focus more on their own branded experience while larger ships will continue to push the limit of what's possible at sea. Below water viewing chambers, on-board observatories, electronic gaming tournament centers, pet spas, cook-your-own dinner-dining rooms, slash-casters, movies-under-the-stars outdoor theaters, graffiti walls, and cruise-for-a-cause walls of fame are just a few of the possibilities here.

In addition, onboard wireless networks will give rise to interactive game playing through personal cellphones. Ship-based photo competitions, audience voting, "complete this phrase" (text in an answer), unusual scavenger hunts, remote ship webcam monitoring, and invite-a-friend games can add entirely new dimensions to what is currently being offered. Some ships may even begin to use flying drones for such things as extended view whale-watching, storm monitoring, cloud formation, and weather analysis. While near land, drones can be used to view the surrounding countryside and even witness city lights at night.

Increases in Multi-Generational Travel

The CLIA fleet carries over 1.6 million kids traveling each year and that number is increasing, in part due to the growth of multi-generational bookings. One recent survey found that 46 percent of families have taken two to four cruises with children under the age of 18, and 15.2 percent have taken five to seven cruises, and 4.8 percent have taken more than ten. As life expectancy grows, and 80-90 year olds become increasingly more active, cruise lines will find themselves needing an even broader selection of programming, with a range of offerings that appeal to even more age groups. Adding to the complexity of age-related programming will be the diverse, rapidly-changing interests of multi-cultural age groups.

Shorter Lead Times

Businesses around the world are beginning to grapple with the fact that shorter lead times are getting shorter every year. Our rapidly accelerating communications networks are constantly raising the bar. The once radical notion that packages and letters could be delivered anywhere in the world overnight, is now stogy thinking, far too slow for today's on-demand generation. For cruise lines, representing an industry built around the leisurely pace of leisure, this creates a number of friction points, as well as several advantages. Customers, who would have booked 6-12 months in advance in the past, now see little need to book more than 1-2 months in advance today. This last-minute thinking that causes heartburn from an operational standpoint also opens the door to last-minute promotional schemes that can insure a near-capacity turnout virtually every time. While switching embarkation and destination points still needs advance planning, it requires far less than it did even a couple years ago. Customer notifications can happen quickly and last minute requests and changes are becoming much more manageable. Onboard staff and talent can be booked with little notice and

Floating Cities and Floating Nation-States

The cruise industry has been quietly testing the limits of international law by asking the fundamental question, "What things can happen in international waters that are not permitted inside most countries?" They are already claiming exemption from sales tax, gambling laws, HR requirements, minimum wage laws, and a multitude of other restrictions that land-based businesses have to deal with. But how far are they willing to push it? And how far is too far?

- Could a medical tourism ship be stationed in international waters to perform medical procedures that are still pending approval in other countries?
- Is it possible for a ship to serve as a shopping center for illicit merchandise such as counterfeit software, illegal arms, human organs, designer drugs, and more?
- Can they create and enforce their own laws, begin to incorporate businesses, develop their own currency, manage their own banking operations, and serve as a tax haven?
- In short, is it possible for a ship to become its own sovereign nation?

If this line of thinking sounds too extreme, consider the following:

1. Floating cities on the ocean have been receiving added attention due, in part, to concerns over climate changes.
2. A number of groups are already in the early development phases of floating and undersea cities of the future.
3. One floating city is already in operation. This super cruise ship off the coast of Florida is operated by Residensea. It has been designed for wealthy families who wish to live and work from a sea-based permanent home floating in international waters.
4. Seasteading Institute in California is based on the belief that "...current political systems are outdated and work poorly, for two reasons. One is the lack of a frontier - a place to go try out new forms of government (like the crazy new "democracy" which sprung up in far-off America). The other is the lack of mobility on land that happens because people are tied to buildings and buildings are fixed in place."
5. SeaLand is a manmade structure 60 miles off the coast of England that has established itself as an autonomous country.

Below are photos of these early efforts:

ResidenSea is a $264 million, 43,000 ton, 12-deck, 644 foot long, 650-person capacity, one-of-a-kind resort floating community at sea that circles the globe. The vessel took six years to build and contains luxury studios, one, two and three-bedroom residences.

Belgian Architect Vincent Callebaut has conceived of a lillypad-designed city that features water collection and purification powered primarily by solar and wind energies.

The Lillypad plan is moving from design to implementation stage, an indication that the development of floating cities may not be so distant.

In Tokyo, to deal with a lack of developable land, a "city in a pyramid" has been proposed to float on Tokyo Bay.

András Gyõrfi's "The Swimming City" was the winner of the Seasteading Institute's first 3D design competition in 2009

Extreme Ship Designs

Cruise lines have proven that they are sitting on top of a very profitable industry and the more outrageous the ship, the more profitable it becomes. This line of thinking is paving the way for a new era of extreme ship designs and extreme operational strategies. I should note that something only sounds extreme before it's built and operating. After it becomes a successful, the label "extreme" gets traded in for "genius."

Here are a few ideas that have been proposed:

Designed by Fredrik Johansson, a mother ship would carry smaller vessels

Waterstudios, Royal Haskoning, and Dutch Docklands have formed a consortium to have this Floating Cruise Terminal built at sea by 2014.

Freedom Ship is a "floating city" concept proposed by Norman Nixon of Freedom Ship International. With a design length of 1400 meters (4500 feet), width of 230 meters (750 feet) and high of 110 meters (350 feet), Freedom Ship is four times the length of the largest cruise ship in the world today.

Addressing the issue of rising sea levels, Russia-based architectural firm Remistudio proposes The Ark, an arch-shaped floating hotel as a refuge from even extreme floods.

Fig. The Ark.

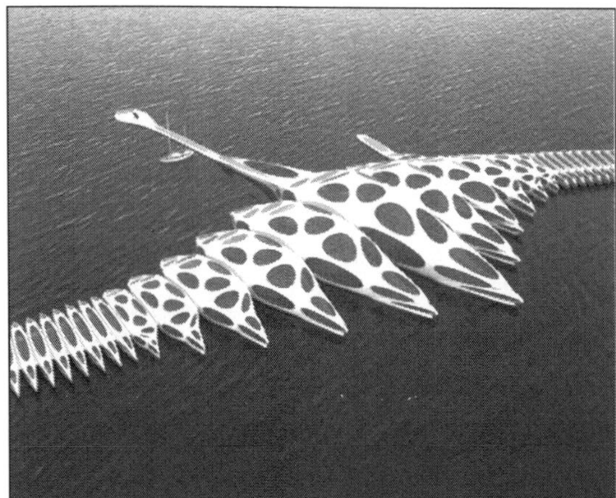

Conceptualized by Gianluca Santosuosso, the MORPHotel is a unique project that intends to develop a new luxury hotel concept where the user can live inside a floating system that keeps moving around the world. The MORPHotel is able to adapt its shape to the weather conditions and the site morphology, thanks to the linear structure developed around the vertebral spine.

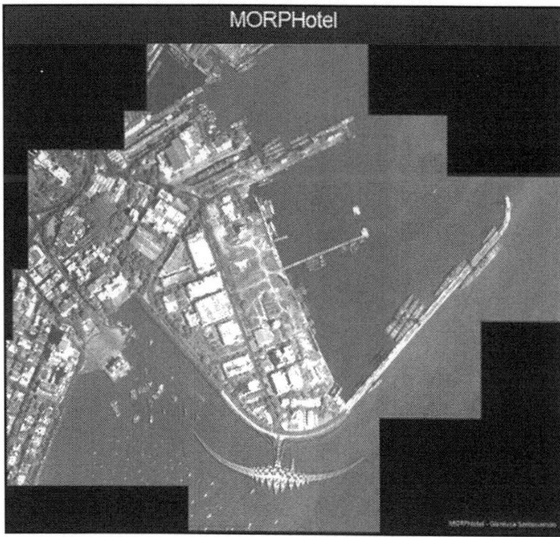

Fig. MORPHotel.

Final Thoughts....

The cruise industry is in the enviable position of having demand build faster than capacity can be created. They can shift assets to match the markets and morph their offerings to match customer demands. And, being outside of most jurisdictional boundaries, they can literally write their own rules.

For this reason they have the ability to manage risks more efficiently than most other industries. But cruise ships still have enormous challenges ahead. For one, the average person gains seven pounds onboard a seven day cruise. If they could figure out a way for the average person to lose seven pound while consuming the same food, then we're talking.

TWO MAJOR FEATURES OF THE CRUISE INDUSTRY

Fundamentally, cruise shipping is a branch of the tourism industry, not one of the transport industry (Dickinson and Vladimir, 1997). It has inherited of the legacy of the great passenger liners, but nowadays the bulk of the ocean-going cruise fleet consists of purpose built vessels. Most of the former liners that had a second lease of life as cruise ships have now been scrapped, are idle or are confined to secondary roles (Middlemiss, 1997), with a few major exceptions like the panamax-sized Queen Elizabeth 2 (actually built as a dual purpose ship, namely a transatlantic liner in summer and a cruise ship in winter). Until recently, this was also the case for the Norway, but this overpanamax-sized vessel is currently idle after a major technical incident in 2003 and her future is unclear.

She spent the first part of her career on the North Atlantic as the luxury passenger liner France and the second part of it as a mass market cruise ship, sailing in the Caribbean for Norwegian Cruise Line (Durand, 2004). Her former American competitor, the United States, is also idle (but in this case since 1969) and she is reported to be in a very bad condition (Miller, 2003). Nevertheless, Norwegian Cruise Line is planning to rebuild her for the US market; as she is panamax-sized, one of her possible uses as a cruise ship under the US flag (after a total rebuilding) is to offer U-shaped cruises between the Atlantic and Pacific US coasts via the Panama Canal.

There is no comprehensive statistical source for the cruise industry and we established therefore our own data base for the analyses offered below, withShipPax's Guide 2003 as the main source of the vessels' technical and ownership details. This unpublished data base reflects the situation as at 01-01-2003 and includes 230 ocean-going cruise ships (excluding purely coastal and river units) with a passenger capacity of at least 100 lower berths. This is a better statistical unit than the overall number of passengers, including the upper berths in some cabins, if any, because the number of such lower berths shows the number of cabins and, hence, the physical size of the ship. It excludes old vessels idle for some time and with little prospects of sailing again (including the above-mentioned United States, but not the Norway that was still in service at that time), the more so as the latest SOLAS rules for safety at sea are coming into force.

The overall number of lower berths of the said 230 cruise ships amounted to 249,630 and their overall gross tonnage to 9,275,050 gt (gross tons). The gross tonnage is also often referred to in cruise shipping in order to characterize the size of a cruise ship, and both it and the number of lower berths will be used here concurrently, as there is no strict correlation between both measurement units. Actually, their quotient, known as the space ratio, is showing the qualitative level of the ship: the higher the number of gross tons per passenger, the higher the level of comfort and of luxury of the ship. For the whole cruise fleet, it amounted to 37.16, but this world average has little meaning, as there are significant differences between cruise lines, and even between ships within the fleet of a given line.

THE OVERWHELMING DOMINANCE OF A FEW MAJOR PLAYERS

In a register of the world cruise fleet, a large number of medium-sized and small cruise lines can be identified, but no real mega player seems to be dominating the market. The largest individual cruise line as at 01-01-2003, namely Royal Caribbean Cruise Line, had only 16 ships in service that is less than 7 per cent of the 230 above-mentioned active vessels. They were accounting for 1,445,100 gt and 37,712 lower berths, and capacity wise this was however 15.6 and 15.1 per cent, respectively, of the two world grand totals for gross tons and lower berths. As can be seen at Table 5, where the fleets in service and on order of the twenty largest individual cruise lines are shown, there was only one line on equal foot with Royal Caribbean Cruise Line, namely Carnival Cruise Lines (it is usually ranking first, actually, and it was due to rank first again in 2004, as it had more ships on order). There was a big gap with lines ranking further down capacity wise, like Princess Cruises, Celebrity Cruises, Holland America Line, Norwegian Cruise Line or Costa Crociere, whose fleets in service accounted for between 750,000 and 400,000 gt, with between 18,000 and 10,000 lower berths each.

Table. The Twenty Largest Individual Cruise Lines (Situation as at 01-01-2003).

	Ships in service			Ships on order		
	Ships -	Gross tonnage	Lower berths	Ships -	Gross tonnage	Lower berths
Royal Caribbean Cruise Line	16	1,445,100	37,712	3	318,500	7,338
Carnival Cruise Lines	18	1,375,600	38,122	4	416,300	11,036
Princess Cruises	10	744,600	17,966	5	550,000	13,322
Celebrity Cruises	9	681,050	16,450	-	-	-
Holland America Line	11	619,900	14,470	3	252,000	5,904
Norwegian Cruise Line	8	599,650	17,002	1	72,000	1,900
Costa Crociere	8	408,600	10,784	3	295,700	7,550
P&O Cruises	4	300,300	7,682	-	-	-
Star Cruises	6	276,500	7,482	-	-	-
Festival Cruises	6	212,600	6,459	-	-	-
Disney Cruise Lines	2	166,700	3,508	-	-	-
Royal Olympia Cruises	11	136,150	5,522	-	-	-
Radisson Seven Seas	5	124,350	2,082	1	46,000	720
Sun Cruises	4	100,300	4,570	-	-	-
Crystal Cruises	2	99,650	1,840	1	68,000	1,096
Cunard Line	2	94,850	2,326	2	226,000	4,362
Louis Cruise Line	7	84,600	3,787	-	-	-
Aida Cruises	2	80,850	1,452	1	42,300	1,266
Mediterranean Shipping Cruises	3	72,050	2,537	2	117,200	3,052
Arosa Cruises	1	69,850	1,590	-	-	-

Computed from ShipPax's Guide 2003 (oceangoing vessels with at least 100 lower berths) But as for container shipping, many of these cruise lines are part of larger corporate groups, for the very same reasons (search for financial and operational economies of scale, as well as of commercial scope), and the above ranking is therefore misleading. The one shown at Table 6 is much more meaningful, with four major world players on the one hand, and a series of independent lines on the other hand. The largest sixteen of these are identified there (with, at that time, only one grouping among these as far as ownership is concerned, namely between Louis Cruise Line and Royal Olympia Cruises, whose alliance came however recently to an end because of the latter's big financial troubles).

Table. The Overwhelming Dominance of Four Large Groups on the World Cruise Market (Situation as at 01-01-2003).

	Ships in service			Ships on order		
	Ships -	Gross tonnage	Lower berths	Ships -	Gross tonnage	Lower berths
Four largest players	*109*	*7,000,900*	*182,366*	*22*	*2,176,800*	*52,678*
Carnival Corporation	46	2,560,650	67,070	12	1,190,000	28,852
Royal Caribbean Group	26	2,166,300	55,674	3	318,500	7,338
P&O Princess Cruises	20	1,341,450	33,238	6	596,300	14,588
Star Goup	17	932,500	26,384	1	72,000	1,900
Independent lines	*121*	*2,274,050*	*67,264*	*5*	*235,200*	*4,868*
Louis/ Royal Olympia Cruises	18	220,750	9,309	-	-	-
Festival Cruises	6	212,600	6,459	-	-	-
Disney Cruise Lines	2	166,700	3,508	-	-	-
Radisson Seven Seas	5	124,350	2,082	1	46,000	720
Sun Cruises	4	100,300	4,570	-	-	-
Crystal Cruises	2	99,650	1,840	1	68,000	1,096
Mediterranean Shipping Cruises	3	72,050	2,537	2	117,200	3,052
Thomson (TO's cruise division)	2	60,400	2,214	-	-	-
Fred Olsen Lines	3	58,950	1,970	-	-	-
Pullmantur	2	58,950	1,696	-	-	-
Hapag-Lloyd	4	58,470	1,162	-	-	-
Phoenix Seereisen	2	53,000	1,732	-	-	-
Nippon Charter Cruises	2	45,224	722	-	-	-
ResidenSea	1	43,188	699	-	-	-
Transocean Tours	2	39,200	1,002	-	-	-
Saga Holidays	2	36,439	955	-	-	-
Other lines	61	823,880	24,807	1	4,000	138
Overall	**230**	**9,275,050**	**249,630**	**27**	**2,412,000**	**57,546**

Computed from ShipPax's Guide 2003 (oceangoing vessels with at least 100 lower berths) The above-mentioned four major players were, as at 01-01-2003, Carnival Corporation (46 ships, 2,560,650 gt and 67,070 lower berths), the Royal Caribbean Group (26 ships, 2,166,300 gt and 55,674 lower berths), P and O Princess Cruises (20 ships, 1,341,450 gt and 33,238 lower berths), and the Star Group (17 ships, 932,500 gt and 26,384 lower berths). Together, they accounted for 109 ships, that is slightly less than half of the fleet, but as there were much bigger than average, their combined shares reached 75.5 per cent for gross tonnage (700,900 gt) and 73 per cent for lower berths (182,366). This is a far higher capitalistic concentration degree than in the container industry (as shown above, the four largest container players amounted to just one third of the TEUs for the cellular fleet and to as little as 25.5 per cent of the overall container fleet).

Moreover, the trend is even towards further concentration. On the one hand, as will be seen below in greater details, a mega merger took place early in 2003 between two of these groups and another merger might take place soon or later between the other two, leading to a quasi duopolistic situation (like in the aircraft industry with Airbus and Boeing). And on the other hand, even without those mergers, their combined shares was anyhow due to be higher in the near future as most of the cruise vessels on order were ordered by these said four groups (with 22 ships on order out 27, for 90 per cent in gross tonnage and 91.5 per cent in lower berths).

Among the medium-sized independent lines, Mediterranean Shipping Cruises (MSC) and Radisson Seven Seas (with, at that time, two vessels on order each) as well as Crystal Cruises (with one vessel) were also due to grow, even if both MSC and Radisson were planning to sell some older, smaller tonnage. Moreover, MSC had the unique opportunity of growing further after buying in 2004 two large ships of the bankrupt Festival Cruises, a much cheaper option than buying out this line as a whole, including the goodwill associated with the brand. MSC has also ordered in 2004 two panamax-sized newbuidings, and its fleet will thus develop dramatically in a very short time; however, the cruise division of the MSC group will remain far smaller than its container division.

THE HIGH VALUE OF CRUISE BRANDS

In the container industry, a lot of mergers took place. On the one hand, this has led to the amalgamation of fleets into a combined brand reflecting the previous individual corporate identities (for example, Maersk Sealand, P and O Nedlloyd, Hanjin / Senator, CMA CGM and Hapag-Lloyd). And on the other hand, one of the two brands has sometimes disappeared (for example, Uniglory into Evergreen, NOL's container division into APL, CMBT into Safmarine). This is because container lines are all selling more or less the same standard product, and brands are therefore of less value than in the cruise shipping industry, where they reflect a strong regional anchorage. Moreover, at least within the largest three groups, they reflect significant qualitative differences (shown for example in the Berlitz Guide to Cruise Shipping by a rating system from one to five stars). However, some subsidiaries have kept until now a distinct identity with large container groups, like Safmarine within AP Moller or Lloyd Triestinoversus Evergreen, and there is one major exception with CP Ships that remains a constellation of individual lines (with some slot exchanges, however).

As can be seen at Table 7, there were, as at 01-01-2004, no less than nineteen individual brands within the four largest groups of cruise lines. In all cases, the main lines are positioned, qualitatively, in the middle of the upper tier of the market, and they have either sister brands at the same qualitative level in another region or distinct brands positioned in another, upper or lower tier of the market.

Table. The Fleet of the Four Largest Cruise Groups and of their Brands as at 01.01.2003.

	Ships in service			Ships on order		
	Ships -	Gross tonnage	Lower berths	Ships -	Gross tonnage	Lower berths
Carnival Corporation	***46***	***2,560,650***	***67,070***	***12***	***1,190,000***	***28,852***
Carnival Cruise Lines	18	1,375,600	38,122	4	416,300	11,036
Costa Crociere	8	408,600	10,784	3	295,700	7,550
Holland America Line	11	619,900	14,470	3	252,000	5,904
Cunard Line	2	94,850	2,326	2	226,000	4,362
Seabourn Cruise Line	3	29,850	612	-	-	-
Windstar Sail Cruises	4	31,850	756	-	-	-
Royal Caribbean Group	***26***	***2,166,300***	***55,674***	***3***	***318,500***	***7,338***
Royal Caribbean Cruise Line	16	1,445,100	37,712	3	318,500	7,338
Celebrity Cruises	9	681,050	16,450	-	-	-
Island Cruises (joint venture)	1	40,150	1,512	-	-	-
P&O Princess Cruises	***20***	***1,341,450***	***33,238***	***6***	***596,300***	***14,588***
Princess Cruises	10	744,600	17,996	5	550,000	13,322
P&O Cruises	4	300,300	7,682		-	
Other British brands (three)	3	145,850	3,538	-	-	-
German brands (two)	3	150,700	4,042	1	42,300	1,266
Star Group	***17***	***932,500***	***26,384***	***1***	***72,000***	***1,900***
Star Cruises	6	276,500	7,482	-	-	-
Norwegian Cruise Line	8	599,650	17,002	1	72,000	1,900
Orient Line	2	56,350	1,900	-	-	-

For example, within the Carnival Cruise group, Carnival Cruise Lines and Costa Crociere are positioned on the same qualitative level, but their main markets are North America and Europe, respectively. Whereas within the Royal Caribbean group, whose main market is North America with Europe as a secondary market, Celebrity Cruises are positioned at a higher level than Royal Caribbean Cruise Line. A combination of the two strategies was characterizing both the Carnival Corporation and P and O Princess Cruises before their recent merger, whereas a merger between the Royal Caribbean and Star groups could be supported by the same combined strategy of brand positioning. In the recent history of cruise shipping, there has been only four exceptions to this general trend of keeping the individual identities of the lines being bought, namely for

Home Lines merged into Holland America Line, for Admiral Cruises merged into Royal Caribbean Cruise Line, for Royal Cruise Line merged into Norwegian Cruise Line, and for Sitmar Cruises merged into Princess Cruises. However other lines have disappeared, either because they were closed down by their owners (such as Royal Viking Line at the time it was a sister company of Norwegian Cruise Line) or because they went into bankruptcy (like Premier Cruises, Renaissance Cruises or more recently Festival Cruises).

Computed from ShipPax's Guide 2003 (oceangoing vessels with at least 100 lower berths) As at 01-01-2003, the Carnival Corporation was the largest of the four major groups, with six well established brands. Carnival Cruise Lines is positioned at the upper tier of the mass market in North America and Costa Crociere is at the same qualitative level in Europe; Holland America Line is a premium line whose main market is North America but with worldwide operations; Cunard Line is a luxury line for North Americans and Europeans and with two niche market operations using mega yacht-sized ships. These are Seabourn Cruise Line for the luxury market and Windstar Sail Cruises for, as its name implies, unconventional cruises under sail, aimed at the premium market; in both cases, most of their guests are drawn from North America. These two niches are marginal, however, and they will be even more in the future, as there were no ships on order for them (but ClubMed 2 could perhaps be bought by Carnival for Windstar, as ClubMed 1, now Wind Surf, has already been), and as there will be a dilution effect with the recent merger between Carnival Corporation and P and O Princess Cruises.

The structure of the Royal Caribbean Group is much simpler, with the main brand (Royal Caribbean Cruise Line) positioned between the mass market and the premium segment, and with another strong brand (Celebrity Cruises) at the upper tier of the said premium market. In both cases, North American passengers are the more numerous, but there is also a new European joint-venture (Island Cruises) for the European mass market. Currently, all of the ships on order are for the main line, whose relative importance within the group will therefore increase further.

Before it merged, early in 2003, with the Carnival Corporation, P and O Princess Cruises were themselves the product of the takeover of the US-based Princess Cruises, positioned on the same qualitative segment as Royal Caribbean Cruise Line, and of the cruise shipping branch of the P and O Group. The latter was featuring a series of brands, each targeted at a specific segment, with the main line (P and O Cruises) positioned qualitatively as Princess Cruises but with a strong British flair, and a series of five highly specific brands. On the one hand, there are three brands for the British or Australian markets, namely Ocean Village (a new club-style operation for British guests), P and O Holidays (mass market cruises for Australians), and Swan Hellenic (cultural cruises for British passengers). And on the other hand, there were at that time two different

German brands (marketed under the umbrella of Seetours), a conventional product (Arosa) and a club ship operation (Aida).

P and O Princess Cruises was the group where the advantages of a group structure were the more visible in the optimal deployment of the fleet, thanks to a series of internal transfers. Two of them took place from Princess Cruises to P and O Cruises (a horizontal transfer, qualitatively speaking), one from Princess Cruises to P and O Holidays (a vertical transfer), one from Princess Cruises to Arosa (a diagonal transfer) and one from Princess Cruises to P and O Cruises, then to Ocean Village (a horizontal followed by a vertical transfer). Another series of transfers took place in 2003 and 2004 after the merger with Carnival Corporation, including reverse transfers, and a reshuffle took place on the German market where Arosa merged into Aida; however, Costa Crociere, now a sister company, has kept under its own umbrella one ship specially aimed at this market, what is strategically questionable in a corporate perspective.

By contrast, within the larger Carnival Corporation empire, such internal moves took place only three times before the said merger, from Carnival Cruise Lines to Costa Crociere (a horizontal transfer), from Holland America Line to Costa Crociere (a vertical transfer) and from Seabourn to Holland America Line (also a vertical transfer).

And except for transferring one ship from the main line to the above-mentioned new European joint venture (horizontally and vertically), this did not happen within the Royal Caribbean Group.

On the contrary, such transfers have been quite common within the fourth major player, namely the Star group, where two new buildings ordered for the Malaysian parent, Star Cruises, were redirected before entering in service to its American subsidiary, Norwegian Cruise Lines, because of the changing conditions on the Asian market. Star's (over)ambitious expansion plans could be elegantly scaled down there thanks to last-minute transfers before the ships were delivered (implying some internal adaptations but no significant extra cost). Moreover, Star's two previous new buildings have recently gone to Norwegian Cruise Line, in one case in the context of an exchange of vessels, showing another example of flexible fleet adaptations within a group of cruise lines.

Even if Star Cruises is the mother company, the main line within the Star Group is currently its US subsidiary, Norwegian Cruise Line. Star Cruises' fleet is more disparate and this is the line where there were the more numerous «come-and-go» in the fleet in the last few years, with only two new buildings in service as at 01-01-2003, and no vessel on order after the second series of two were actually delivered to Norwegian Cruise Line. There is also a third line in the group, Orient Line, aimed at North American and European guests, a two-ship operation as at the end of 2002 but with one of the two vessels going back thereafter to Norwegian Cruise Line in another example of reverse horizontal transfer. This group is the only one with a new brand currently in the pipeline, to be called NCL America, with US-flagged and US-crewed vessels to operate on the Hawaii market after the bankruptcy of American Hawaii Cruise Lines, with a fleet including up to two reflaged existing foreign vessels, two new buildings ordered by a previous, now bankrupt operator and perhaps the former transatlantic liner United States, if she undergoes a total rebuilding after being laid-up for thirty-five years.

CARNIVAL CORPORATION/PLC AS THE NEW INDUSTRY LEADER

As mentioned above, the first and third largest players in the cruise industry merged in April 2003.Technically, it has taken the form of a takeover of P and O Princess Cruises by Carnival Corporation through the formula of a dual-listed company, with Carnival Corporation quoted in New York on the NYSE and Carnival plc quoted in London on the LSE (with about 80 and 20 per cent, respectively, of the market capitalization).

Table. From Four to Three (then Two?) Large Cruise Groups (Situation as at 01-01-2003).

	Ships in service			Ships on order		
	Ships -	Gross tonnage	Lower berths	Ships -	Gross tonnage	Lower berths
Situation at 01-01-2003						
Carnival Corporation	46	2,560,650	67,070	12	1,190,000	28,852
Royal Caribbean Group	26	2,166,300	55,674	3	318,500	7,338
P&O Princess Cruises	20	1,341,450	33,238	6	592,300	14,588
Star Group	17	932,500	26,384	1	72,000	1,900
Initial merger proposal						
Royal Caribbean / P&OPC	46	3,507,750	88,912	9	910,800	21,976
Carnival Corporation	46	2,560,650	67,070	12	1,190,000	28,852
Star Group	17	932,500	26,384	1	72,000	1,900
Possible countermerger (1)						
Royal Caribbean / P&OPC	46	3,507,750	88,912	9	910,800	21,976
Carnival / Star	63	3,466,285	93,454	13	1,262,000	30,752
Current merger						
Carnival Corporation / plc	66	3,902,100	100,308	18	1,782,300	43,440
Royal Caribbean Group	26	2,166,300	55,674	3	318,500	7,338
Star Group	17	932,500	26,384	1	72,000	1,900
Possible countermerger (2)						
Carnival Corporation / plc	66	3,902,100	100,308	18	1,782,300	43,440
Royal Caribbean / Star	43	3,098,800	82,058	4	390,500	9,238

The impressive domination of the new giant can be seen at Table 8, where their combined fleets amounted, as retroactively at 01-01-2003, to 66 vessels (that is 28.5 per cent of the world fleet), with 3,902,100 gt and 100,308 lower berths (that is, respectively, 42.0 and 40.2 per cent of the world grand totals). The said domination will be even higher in the near future, as most of the ships on order (18, for another 1,782,300 gt and 43,300 lower berths) will go to the lines of a new giant sometimes referred to as Carnivore Cruise Lines. This should be compared to the 13 per cent of the cellular fleet and the 10 per cent of the whole container fleet shown above for the container industry leader, the AP Moller Group.

For many industry observers, it has been hard to understand how the US, European and British antitrust authorities gave their green light to such a mega merger, the more so as Carnival Corporation/plc will find itself in a situation of absolute domination on some key markets. This has been especially the case in Alaskan waters, with 14 ships for four of its brands in the 2003 summer season (for 847,000 gt and 20,094 lower berths) against only 10 ships (excluding coastal vessels) for five competing lines (for 724,650 gt and 17,475 lower berths). This even less understandable as Holland America Line and Princess Cruises are there in an overwhelming position ashore (at least 80 per cent) thanks to their respective subsidiaries Westours and Princess Tours. These offer highly popular packages combining cruises and pre- or post-cruise tours in Alaska, including by train, for which they own the whole fleet of panoramic railroad equipment. Strangely enough, Carnival Corporation/plc has been given the green line without even being obliged to disinvest somewhat in Alaska in order to keep there a reasonable degree of competition. One of the reasons given by the antitrust bodies (and by Carnival/Princess in their application) is that the main competition is not between individual cruise lines or cruise groups, but between the cruise industry as a whole and other forms of tourism. It remains to be seen if they were right and if Carnival Corporation/plc will be tempted in the future to abuse of its dominant position.

Actually, the initial merger proposal came from the Royal Caribbean Group and it had first the favour of P and O Princess Cruises' management. It would have lead to a much more balanced situation, with 46 ships each for Carnival Corporation in its original definition and for the Royal Caribbean / P and O Princess combination.

This would have ranked first, with 3,507,750 gt and 88,912 lower berths, against 2,560,650 gt and 67,070 lower berths for Carnival Corporation, but the imbalance could have been easily rectified as Carnival had more capacity on order and as it had the financial means available to merge with another player. For example, a merger with the Star Group would have lead with a quasi perfectly balanced duopolistic situation, capacity wise (3,466,300 and 93,454 lower berths for a Carnival / Star combination, against 3,507,750 gt and 88,912 lower berths for the Royal Caribbean / P and O Princess combination). To counterbalance the new Carnival Corporation/plc galaxy, the only solution left in the short term is a Royal Caribbean / Star combination, but such a merger would lead to a slightly less powerful counterweight in the short term (3,098,800 gt and 82,058 lower berths) and to an increased gap after the ships currently on order are delivered.

Such a situation would have never been allowed to happen in the container shipping or in the airline industries, and one still wonder why, at least, some significant divestures were not even asked for, especially by the US authorities. For example, breaking P and O Princess Cruises into two parts, with only its European brands joining Carnival Corporation and Princess Cruises being offered for sale to Star Cruises, would have led to a much more balanced situation. There would have been then three large players on more or less equal foot and with no major qualitative and geographic overlap between their brands, as it is now the case with Holland America Line and Princess Cruises within Carnival Corporation/plc (not only in Alaska in summertime, but also in Europe at that time and in the Caribbean in winter).

THE INCREASING SHARE OF PANAMAX AND OVERPANAMAX CRUISE SHIPS

The cruise industry has entered into the panamax and overpanamax eras, for the very same reasons (search for economies of scale) but much later than container shipping.

Table. Structure of the World Cruise Fleet in Service as at 01-01-2003 by Gross Tonnage Classes.

	Carnival Corporation	Royal Caribbean	P&O Princess	Star Group	Other lines	Grand total
Infrapanamax	***19***	***3***	***7***	***10***	***119***	***158***
< 20,000	7	-	-	1	75	83
20,000 – 39,999	6	-	4	4	37	51
40,000 – 59,999	6	3	3	5	7	24
Panamax	***23***	***19***	***10***	***6***	***2***	***60***
< 60,000	5	1	-	-	-	6
60,000 – 79,999	13	12	9	4	2	40
80,000 – 99,999	5	6	1	2	-	14
Overpanamax	***4***	***4***	***3***	***1***	-	***12***
< 100,000	-	-	-	1	-	1
100,000 – 119,999	4	-	3	-	-	7
120,000 – 139,999	-	4	-	-	-	4
≥ 140,000	-	-	-	-	-	-
Overall	**46**	**26**	**20**	**17**	**121**	**230**

As Table 9 shows, out of the 230 cruise vessels in service as at 01-01-2003, 157 were infrapanamax-sized (namely with a beam of less than 100 feet or 30.5 meters, as defined by the Panama Canal Authority). However, there were 61 panamax-sized vessels (with a beam up to 32.3 meters) and already 12 were overpanamax-sized (with a beam above 32.3 meters, currently up to 41.0 meters for Cunard's Queen Mary 2 delivered at the end of 2003). In the latter two categories, two special, older ships built as transatlantic liners are included, namely Cunard's Queen Elizabeth 2 on the one hand and Norwegian Cruise Line's Norway on the other hand (currently idle after a technical incident). Both were built in the 1960s, whereas all the other large and very large cruise ships were purely built for cruising, as many of the smaller vessels but much later than these. This is simply because it would have been commercially impossible to fill ships with more than, say, 2,000 passengers before cruise shipping had become a mature industry with a rather large customer base, first on the North American market, then in Europe and more globally.

As can be seen at Figure 2, the first purpose-built panamax-sized cruise ship (Royal Caribbean Cruise Line's Sovereign of the Seas) entered in service only in 1987, and this segment of the cruise fleet grew steadily since then, with a first maxipanamax (a special category, for which the length is the maximum accepted in the Panamax locks, namely 294.0 meters, as this was already the case for Queen Elizabeth 2) joining the fleet in 1998 (Disney Cruise Lines'Disney Magic). Actually, the latter came into service after the first purpose-built overpanamax, which entered in service in 1996 (Carnival Cruise Lines' Carnival Destiny), as cruise lines saw first the benefits of going overpanamax in terms of economies of scale. However, they realized quickly that they also needed, for part of their fleet, the flexibility offered by the panamax design, optimized nowadays to more than 90,000 gt and perhaps up to 100,000 gt in the Finnish Nova project). Moreover, very few standard panamax cruise ships were built recently, and only one was on order in early 2003 (four more followed since that time, however); this is also the case for container shipping where the panamax design has been gradually improved to vessels of more than 4,000 TEUs of maxipanamax standard (also known as panamax-max).

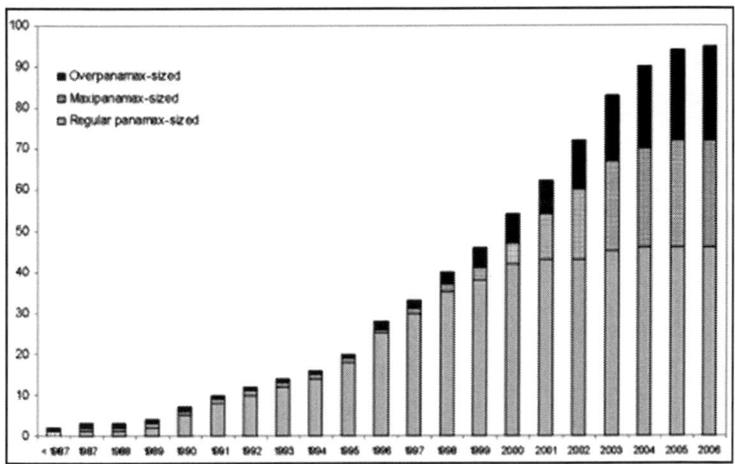

Fig. The Rise of the Panamax- and Overpanamax-Sized Cruise Fleet (Number of Ships).

What is unique to the cruise industry is that virtually all of the large or very large cruise ships in service (Table 9) or on order (Table 10) are or will be operated by lines part of the four above-mentioned dominant groups (three nowadays, after the Carnival/Princess merger). As at 01-01-2003, there were

23 panamaxes and 4 overpanamaxes in service within the Carnival Corporation fleet (plus another 6 and 6 on order), and the figures were 19 + 4 (and 2 + 1) for the Royal Caribbean Group, 10 + 3 (and 1 + 4) for the then independent P and O Princess Cruises, and 6 + 1 (but only 1 + 0) for the Star Group (excluding the planned conversion of the maxipanamax-sized United States). The latter group was clearly behind its rivals in the recent search for economies of scale, and its fleet still featured a majority of infrapanamax-sized vessels (10 out of 17, including three ships that were lengthened in order to be more productive).

On the contrary, infrapanamaxes were already less numerous than the other two, larger categories in the fleets in service of Carnival Corporation (19 out of 46, including 7 very small ships for its two niche market brands); this was also the case for the Royal Caribbean Group (only 3 left out of 26, after most of its smaller tonnage was sold) and for P and O Princess (7 out of 20). At that time, the four major players have only one infrapanamax-sized ship on order (for P and O Princess Cruises' German brand Aida), showing how far they have gone into the race for ever bigger ships. However, they still have the possibility to turn to the second hand market when they need smaller vessels for specific niche markets, as was the case recently for three of the former Renaissance mid-sized ships, two of which went to Princess Cruises and one to Swan Hellenic.

Table. Structure of the World Cruise Fleet on Order as at 01-01-2003 by Gross Tonnage Classes.

	Carnival Corporation	Royal Caribbean	P&O Princess	Star Group	Other lines	Grand total
Infrapanamax	-	-	*1*	-	*4*	*5*
< 20,000	-	-	-	-	1	1
20,000 - 39,999	-	-	-	-	-	-
40,000 - 59,999	-	-	1	-	3	4
Panamax	*6*	*2*	*1*	*1*	*1*	*11*
< 60,000	-	-	-	-	-	-
60,000 - 79,999	-	-	-	1	1	2
80,000 - 99,999	6	2	1	-	-	9
Overpanamax	*6*	*1*	*4*	-	-	*11*
< 100,000	-	-	-	-	-	-
100,000 - 119,999	5	-	4	-	-	9
120,000 - 139,999	-	1	-	-	-	1
≥ 140,000	1	-	-	-	-	1
Overall	**12**	**3**	**6**	**17**	**5**	**27**

OVERPANAMAX-SIZED SHIPS' OWNERSHIP

When going into the detailed structure of the four groups by brand, one sees at Table 11 that, in early 2003, overpanamaxes were operated only by the largest brand in each group, namely Carnival Cruise Lines for Carnival

Corporation, Royal Caribbean Cruise Lines for the Royal Caribbean Group, Princess Cruises for P and O Princess Cruises and Norwegian Cruise Line for the Star Group. This will not change in the near future for the latter three groups (as Star Cruises has dropped its plans to enter into the overpanamax era), but overpanamax-sized vessels will be spreading within the Carnival Corporation (now Corporation/ plc).This will be the case at Costa Crociere (with carbon çopies, except for the funnels, of Carnival Cruise Lines' giants, in a clear search for intragroup economies of scope) and at Cunard Line, with the already mentioned Queen Mary 2.

Table. Structure of the Fleets of the Four Largest Cruise Groups as at 01-01-2003 by Brand and Ship Size.

	Ships in service			Ships on order		
	Infra-panamax	Panamax -	Over-panamax	Infra-panamax	Panamax -	Over-panamax
Carnival Corporation	***19***	***23***	***4***	-	***6***	***6***
Carnival Cruise Lines	3	11	4	-	1	3
Costa Crociere	6	2	-	-	1	2
Holland America Line	2	9	-	-	3	-
Cunard Line	1	1	-	-	1	1
Seabourn Cruise Line	3	-	-	-	-	-
Windstar Sail Cruises	4	-	-	-	-	-
Royal Caribbean Group	***3***	***19***	***4***	-	***2***	***1***
Royal Caribbean Cruise Line	-	12	4	-	2	1
Celebrity Cruises	2	7	-	-	-	-
Island Cruises (JV)	1	-	-	-	-	-
P&O Princess Cruises	***7***	***10***	***3***	-	***1***	***4***
Princess Cruises	3	4	3	-	1	4
P&O Cruises	-	4	-	-	-	-
Other British brands (three)	2	1	-	-	-	-
German brands (two)	2	1	-	1	-	-
Star Group	***10***	***5***	***1***	-	***1***	-
Star Cruises	4	3	1	-	1	-
Norwegian Cruise Line	4	2	-	-	-	-
Orient Line	2	-	-	-	-	-

Computed from ShipPax's Guide 2003

At 148,528 gt, the QM2 is currently the largest cruise ship in service at 148 528 gt. This will not last for long, however, as Royal Caribbean Cruise Line has ordered recently a 158,000 gt mega ship (3,600 lower berths) for delivery in 2006 (with another in option, and probably more in the pipeline). Moreover, Carnival Corporation/plc is said to want to keep the leadership with a planned series of 180,000 giants for Carnival Cruise Lines and for Princess Cruises (sharing a same hull but with different standards of accommodations to reflect the different qualitative positioning of the two cruise lines). Excluding these yet to be confirmed plans and planned repeat orders of current, smaller overpanamaxes for the same two lines, Table 12 shows the state of the overpanamax-sized cruise fleet as at 01-07-2004.

Table. Overpanamax–Sized Cruise Ships Delivered or on Order by Cruise Line as at 01-07-2004.

	Year built	Gross tonnage	Length (m)	Beam (m)	Lower berths
Carnival Cruise Lines					
Carnival Destiny	1996	101,350	272.2	35.5	2,642
Carnival Triumph	1999	101,500	272.2	35.5	2,758
Carnival Victory	2000	101,500	272.2	35.5	2,640
Carnival Conquest	2002	110,200	290.2	35.5	2,974
Carnival Glory	2003	110,200	290.2	35.5	2,974
Carnival Valor	2004	110,200	290.2	35.5	2,974
Carnival Liberty	2005	110,200	290.2	35.5	2,974
Costa Crociere					
Costa Fortuna	2003	105,000	272.2	35.5	2,718
Costa Magica	2004	105,000	272.2	35.5	2,718
Costa TBN (to be named)	2006	110,200	290.2	35.5	3,004
Cunard Line					
Queen Mary 2	2003	148,500	345.0	41.0	2,514
Princess Cruises					
Grand Princess	1998	108,800	289.0	36.0	2,600
Golden Princess	2001	108,800	289.0	36.0	2,592
Star Princess	2002	108,800	289.0	36.0	2,592
Caribbean Princess	2004	120,000	289.0	36.0	3,000
Sapphire Princess	2004	113,000	290.0	37.5	2,674
Diamond Princess	2005	113,000	290.0	37.5	2,674
Crown (?) Princess	2006	120,000	289.0	36.0	3,000
Royal Caribbean Cruise Line					
Voyager of the Seas	1999	137,300	311.1	38.6	3,138
Explorer of the Seas	2000	137,300	311.1	38.6	3,114
Adventure of the Seas	2002	137,300	311.1	38.6	3,138
Navigator of the Seas	2002	138,300	311.1	38.6	3,138
Mariner of the Seas	2003	138,300	311.1	38.6	3,138
UltraVoyager (project name)	2006	158,000	339.0	38.6	3,600
Norwegian Cruise Line					
Norway (currently idle)	1961	76,050	315.5	33.7	2,122

To a much higher degree than for the container industry, there is a clear trend for the major lines to order long series of sister ships or of nearly sister ships for the same brand or across brands, horizontally (Carnival/Costa) or even vertically (Carnival/Princess). Currently, the longest series of identical hulls has been for eight standard panamaxes for Carnival Cruise Lines. However, when all ships on order will be delivered, the longest series will be one of ten overpanamaxes for the same line (seven) and for Costa Crociere (three); a figure that could even extend to twelve or thirteen (another two for Carnival, plus one for Costa).

There is also a current series of five large overpanamaxes for Royal Caribbean Cruise Line (plus at least two in the pipeline, as the Ultra Voyagers are lengthened versions) and another one of five smaller overpanamaxes for Princess Cruises (another two have been ordered recently at the same yard, whereas two rather similar ships with a somewhat different hull, built in another yard; are not included). And for maxipanamaxes, the same trend is being observed with a series of six vessels either for Carnival Cruise Lines (four) or for Costa Crociere (two), with a series of five ships for Holland America Line and with two different series of four identical vessels for Royal Caribbean Cruise Line and for Norwegian Cruise Line (in this case, with two initial orders for Star Cruises transferred to NCL before delivery and a recent order of two quasi sisters directly for NCL).

Except for two overpanamaxes of a different design built for Princess Cruises in Japan (where a few infrapanamaxes have also been built), all these large cruise ships have been built or are on order at European yards. These have managed to keep a dominant position for these sophisticated vessels, whereas most of the containerships are currently built in Asia (except for some of Maersk Sealand ships, built at AP Moller's Danish yard), be it in Japan or in South Korea, whereas Chinese yards are also planning to enter into the race. The European leading builders are Fincantieri in Italy, Kvarner Masa-Yards in Finland, Alstom / Chantiers de l'Atlantique in France and Meyer Werft in Germany, but in the latter two cases the order books are not as strong as what they were previously, and some consolidation might arise as a consequence. Only Fincantieri has a solid order book, but all the ships it will deliver in the next few years will go to the new giant Carnival Corporation/plc, incorporating P and O Princess Cruises. In the near future, this mega customer could try to split its orders between two or three shipbuilders in order to introduce competition between its suppliers. It remains to be seen if, like this has already happened for the cruise groups, a concentration process will not also take place soon in the European shipbuilding industry. This would also lead to an even smaller number of players on the shipbuilding side, to counterbalance the likely rise of Asian shipbuilders.

OVERPANAMAX-SIZED SHIPS' OPERATIONAL CONSTRAINTS

The main reason for the cruise lines to refrain, until the mid-1990s, from ordering overpanamax cruise ships is that they saw total operational flexibility on a seasonal basis as a key criterion of their success. The main geographic complementarity was (and is still) between the Caribbean basin in wintertime and Alaska in summertime (Miller and Grazier, 2004), but there are also two minor, but highly lucrative markets requiring panamax-sized ships, transcanal cruises on the one hand (Charlier, 2000b) and round the world cruises on the other hand (McCalla and Charlier, 2004).

Like Norway since the early 1980s, the first new overpanamaxes started their commercial life by staying year-round in the Caribbean, but their owners found thereafter new geographic and seasonal complementarities for their very large vessels on each side on the Panama canal, the current pattern of which is shown at Table 13. On the Atlantic side, some ships spend or will spend the summer season either in Europe (mainly the Mediterranean, but also Northern Europe) or in the North East Atlantic out of New York (Klein, 2003) before coming back in the Caribbean for winter, whereas on the Pacific side they are deployed in Alaska in summer and spend wintertime either on the Mexican

Riviera or in Asia, after offering a transpacific cruise (Douglas and Douglas, 2004).

Table. Overpanamax–Sized Cruise Ships Cruise Areas in 2004-2005.

	Summer season 2004	Winter season 2004-2005
Carnival Cruise Lines		
Carnival Destiny	Caribbean	Caribbean
Carnival Triumph	Caribbean	Caribbean
Carnival Victory	NE Atlantic	Caribbean
Carnival Conquest	Caribbean	Caribbean
Carnival Glory	Caribbean	Caribbean
Carnival Valor	Not delivered yet	Caribbean
Costa Crociere		
Costa Fortuna	Mediterranean	Mediterranean
Costa Magica	Not delivered yet	Mediterranean
Cunard Line		
Queen Mary	Transatlantic (mainly)	Worldwide
Princess Cruises		
Grand Princess	NW Europe/NE Atlantic	Caribbean
Golden Princess	Caribbean	Caribbean
Star Princess	Mediterranean	Caribbean
Caribbean Princess	Caribbean	Caribbean
Sapphire Princess	Alaska	Mexican Riviera
Diamond Princess	Alaska	Asia / Pacific
Royal Caribbean Cruise Line		
Voyager of the Seas	Caribbean	Caribbean
Explorer of the Seas	Caribbean	Caribbean
Adventure of the Seas	Caribbean	Caribbean
Navigator of the Seas	Caribbean	Caribbean
Mariner of the Seas	Caribbean	Caribbean

This situation will last for at least ten years, before a new set of larger locks are eventually built to improve dramatically the accessibility of the Panama Canal, as planned currently by the Panama Canal Authority. These new locks will be conceived to accommodate very large bulkers or container ships, and there should be no problem for cruise ships up to 200,000 gt to be able to transit this artificial waterway. At least as far as their length, their beam and their water draft are concerned, but the limiting factor will be their air draft, because of the two bridges built across the canal. Therefore, the future generation of panamax cruise ships will be longer and wider than the current overpanamaxes, but these vessels will not be higher on water. And in order to minimize the number of less popular inside cabins, they will feature internal atria (like already for Royal Caribbean Cruise Line's giants) or even open pool or garden areas between two blocks of balcony cabins on each side.

In the meantime, new operational patterns will be explored by current overpanamax owners, the first of which being year-round cruising in the Mediterranean by keeping there their mega ships in winter (Ridolfi, 2000). At least two of the three overpanamaxes of Costa Crociere are due to stay year-round in the Mediterranean. And if it keeps growing at the same pace as in the last five years, it is likely that Mediterranean Shipping Cruises will be the first independant line to enter into the overpanamax era and that they will keep such ships year-round in the Mediterranean, as their non European base is too narrow. The Asia Pacific area, especially South East Asia, whose forthcoming rise makes little doubt (Douglas and Douglas, 2004), is another likely operational theatre for overpamanaxes, and this has already been explored seasonally by Princess Cruises, in tandem with Alaska or Europe. However, cruising there year-round is a more logical option for a regionally based line like Star Cruises, with new Asian-tailored tonnage if, as the regional economic climate is now better, they consider again an earlier project of overpanamaxes.

3

Identify Major Cruise Ports in Each Sector

EVOLUTION OF PORT INSTITUTIONAL FRAMEWORKS

Private sector investment and involvement in ports emerged as a significant issue in the 1980s. By this time, many ports had become bottlenecks to the efficient distribution chains of which they are an essential component. Three main problems, illustrated by port congestion and consequent chronic service failures, contributed to the gradual deterioration of service quality during this period.

The first problem was restrictive labor practices. Increasingly after World War II, antiquated work practices and methods for matching available labor with occasional work practices that developed during a previous era characterized by breakbulk cargo handling needed to be transformed and renegotiated to adjust to modern bulk handling methods, unitized handling, and containerization. All of these developments resulted in a rapid modernization of port handling equipment. At the start of this process, labor unions often refused to accept reductions in the labor force and ignored the need to upgrade skills. Later, however, unions realized that port reform was a necessity. Enlightened labor leaders accepted moderate reforms. As Module 7 describes in greater detail, it is no longer realistic for dock workers and their trade unions to oppose institutional reform and the technological advances that frequently precede and accompany it.

The second reason why many ports failed to respond adequately to the increased demands imposed on them was centralized government control in the port sector. Particularly between 1960 and 1980, central planning (in the port sector as well as in other sectors) prevailed not only as a norm in socialist economies, but also in many western and developing countries where national port authorities were often promoted by international development banks. Slow-paced and rigidly hierarchical planning, control, and command structures often accompanied central planning. Only in the 1980s did the dismantling of communist systems and the increasing introduction of market-oriented policies on a worldwide basis open the way for decentralized port management and for reduced government intervention in port affairs. The third reason for a lack of port service quality was the inability or unwillingness of many governments to invest in expensive port infrastructure or the 'misinvestment' in infrastructure (providing facilities that were badly matched with the needs of foreign trade and shipping). During this period, a number of beautifully constructed port complexes became 'white elephants' when expected demand failed to materialize (see Box 1). As a result of systemic failures in managing port development, governments have learned to rely increasingly on private investors to reduce ports reliance on state budgets and to spread investment risks through joint undertakings.

During this period, fundamental questions arose about the appropriate division of responsibilities between the public and private sectors. 'Boundary line' issues came into sharp focus during the 1980s and 1990s. Policy makers became increasingly aware of the need for coordination among various branches of government and for consultation with diverse port interests. They realized clearly that port development had collateral consequences and effects on public interests in land use, environmental impact, job creation, and economic stimulation for economically blighted areas. Moreover, among some leaders, first in the United Kingdom and then gradually in other parts of the world, it became increasingly clear that large-scale government involvement in port operations was self-defeating and destructive of private initiative. They came to realize that the role of government in a market economy should focus on the provision of public goods (goods and services that the private sector has no adequate incentive to provide and, consequently, are undersupplied without some form of government intervention).

In many countries today, still another trend has emerged: the private provision of public services. Increasingly, governments have transferred public tasks to private contractors. Outsourcing of key functions and roles has had a major impact on redrawing traditional boundary lines in the port sector. Hence, in many ports today, the public sector mainly acts as planner, facilitator, developer, and regulator while providing connectivity to the hinterland, whereas the private sector acts as service provider, operator, and sometimes also developer. Experimentation in shifting the boundary line that divides the public and private sectors has resulted in a healthy pragmatism.

Today, best practice is more concerned with results than with ideology, and is intended to achieve:

- Increased service levels for infrastructure users.
- Increased efficiency in operations.
- Improved allocation of limited public funds.

At the same time, various types of port terminals have become highly specialized in the cargo handling services they provide and manifest fewer of the characteristics of a public good. New greenfield container terminals have been built with private capital, and other container terminals have been redeveloped and recapitalized through some form of private sector participation. Box 2 presents two of the institutional formats used in recent years to develop greenfield terminals.

Increasingly, ports are being integrated into global logistics chains, and the public benefits they provide are taking on regional and global attributes. The value of services provided by regional ports increasingly transcends the interests of local users, and benefits businesses and communities located beyond regional and national borders. This global diffusion of benefits poses some interesting challenges with respect to the need for large-scale investments in the sector. At the same time, as discussed in Module 2, private port service providers themselves have become increasingly global in scope and scale. Even more recently, a number of strategic alliances have formed both within the global shipping industry and the port services industry.

These alliances have profound implications for the ways ports are financed, regulated, and operated. Confronted with these global shipping and port service powers, port authorities will increasingly have challenges in defending public and local interests. Container terminal operators with global coverage, sometimes in alliance with major shipping lines, may be tempted to take advantage of their dominant position to strengthen their network, thereby reducing the scope of competition mainly at the expense of public interests. Moreover, countervailing powers at an international level that have not yet emerged are expected to do so soon due to the absence of suitable national regulating structures. At port level, a strict organizational separation of the commercial and regulating tasks of port management is required to safeguard public interests.

PORT FUNCTIONS, SERVICES, AND ADMINISTRATION MODELS

Ports produce a combination of public and private goods. Public goods include those that are inherently non-divisible and non-consumable, such as public safety, security, and a healthy environment on the one hand, and coastal protection works necessary to create port basins on the other hand. Private goods are both consumable and divisible and their use entails a minimum of economic externalities. Most of the value of private goods can be captured in market transactions between private parties. However, a substantial portion of the value of public goods cannot be captured in arms-length transactions. Consequently, private firms have little incentive to produce them. Public goods create positive externalities when they are used; the social benefits they generate are greater than the price that private parties can charge for them. Thus, some form of public intervention is appropriate in their production to make certain that an adequate level of public goods is maintained.

Ports represent a mix of public and private goods. They generate direct economic benefits (private goods) through their operations, as well as additional indirect benefits (public goods) in the form of trade enhancement, second order increases in production volumes, and collateral increases in trade-related services. These economic multiplier effects have been used by many ports to justify direct public sector investment. It is in this dual production of both public and private goods that complexities arise, which makes defining roles for and boundaries between the public and private sectors challenging in the ports industry. This is particularly the case in the fields of marine and port safety, port security, and the protection of the marine environment. Box 3 lists a number of areas where ports generate economic multiplier effects.

Both through targeted development policies and the unplanned growth of interrelated industries, many ports have become the location for industrial clusters. Industrial clusters are geographic concentrations of private companies that may compete with one another or complement each other as customers and suppliers in specialized areas of production and distribution. Industrial clusters represent a kind of value chain, a web of interrelated activities that are mutually supportive and continuously growing. Clustering of related activities improves the competitive advantage of cluster participants by increasing their productivity, reducing transaction costs among them, driving technological innovation, and stimulating the formation of new business spin-offs.

Large ports offer particularly attractive locations for seed industries and distribution-intensive enterprises. Several notable port-centered industrial clusters have developed over the last 50 years, for instance, those in Dubai, Colon, Norfolk, Rotterdam, Yokohama, Antwerp, Hamburg, Marseilles, and Houston, to name but a few. From the 1950s, the larger European ports targeted refineries and chemical industries for colocation and codevelopment, with considerable success. Thus, for example, a large cluster of five refineries and many chemical-processing companies located in the Port of Rotterdam as a direct result of public policies developed in 1950s. A cluster of world-class, specialized marine services likewise established themselves in the Port of Rotterdam as a result of the good hinterland connections and the gas and oil finds in the North Sea. Another example of cluster development is the Port of Colombo; a fashion goods and apparel industry cluster has developed around Colombo, which focuses on reliable, short-transit container services to complete just-in-time (JIT) purchase orders. This development was business-driven and not the direct result of explicit public policy. The lesson demonstrated in Colombo is that quasi-public goods in the form of efficient industrial networks can be created and developed through private initiatives.

As a matter of strategic development policy, many ports encourage the codevelopment of various value-added services through franchising, licensing, and incentive leasing. Today, ports seek to attract enterprises that extend their logistics chains or provide them with specialized capabilities to add value to cargoes that are stored and handled in the port. General services that many ports attempt to develop include chandlering, ship repair, container maintenance, marine appraisals, insurance claims inspections, and banking. Box 4 describes the efforts of one port to expand and develop its ensemble of value-added services.

Many governments are directly or indirectly involved in port development. They often use a 'growth pole' argument to justify the direct financing of basic port infrastructure. This growth pole rationale derives from the belief that investments in port assets have strong direct and indirect multiplier effects on the entire national economy and, further, that the commitment of public resources is necessary to encourage coinvestment by the commercial and industrial sectors. These sectors are thus stimulated to make investments that they would not make in the absence of public seed investment in port infrastructure. However, determining causal links between public investment and specific commercial activities and investments is difficult and at times speculative. Still, it is important that governments envision and articulate future development scenarios, maintain frequent consultation with the private sector, and implement public policies that are applied consistently and that enable the private sector to invest with confidence in projects that support the stated public policy objectives.

On the other hand, port operations are businesses in their own right and should be managed to achieve optimal utilization of capital. Investments in port assets are affected by risk, competition for land and capital, or other factors in the competitive business environment. Subsidies and government-provided incentives distort the allocation of resources for port development and may result in over- or underinvestment. It is the delicate alignment of public and private interests that determines the structure of port management and port development policy. A full spectrum of institutional frameworks is available, differing primarily in where the boundary line is drawn between the public and private sectors. At one end of this spectrum, full public control over planning, regulation, and operations results in a 'service port.' At the other end, the almost total absence of public ownership, control, or regulatory oversight results in a 'fully privatized port.'

In a clear trend, the alignment of public and private interests in recent years has resulted in a diminishing role for governments in the port industry. The total absence of public involvement in the port sector, however, still remains an exception, limited primarily to specialized ports and terminals. When governments attempt to increase national economic welfare through port development, they may choose to apply one of two distinct normative frameworks: the market surrogate framework or the public interest framework.

In seeking to increase economic welfare, governments may attempt to remedy market imperfections and capture non-market externalities within appropriately engineered and contested transactions. Alternatively, they may pursue explicit goals developed through public consultative processes designed to determine demand for public goods.

With respect to the market surrogate framework, the primary task of government is to identify and eliminate market imperfections and anticompetitive behaviour or to regulate its undesired effects. For example, competition 'for the market' can replace competition 'in the market,' and competition 'for the market' can be engineered into contestable offers of rights in ways that assure procompetitive outcomes. It follows that one of the objectives of public policy should be to create contestable market structures for port services and to manage competitive behaviour. This might be accomplished through licensing, leasing, concessioning, or other methods designed to bring about an efficient allocation of resources. The market surrogate view is followed in most countries with market-oriented economic policies.

The need for some form of government intervention in markets for port services is related to the unique economic characteristics of seaports, some of which tend to make them natural monopolies:

- The provision of port services entails large fixed costs and low marginal costs. The marginal benefits associated with using port services exceed the marginal costs of providing these services.
- A relatively large, minimum initial capacity of basic infrastructure is required for technical reasons.
- The infrastructure is frequently indivisible and, as a result, increases in infrastructure capacity can only be realized in 'quantum chunks.'
- Both initial construction and port expansion require large amounts of capital. As a result, the need to develop basic port infrastructure (for example, sea locks, breakwaters, quay walls, and main roads) all at one time creates large capital operating losses and foregone investment opportunities as a result of underused capacity during the earlier phases of a project's life cycle.
- The life span of port infrastructure projects often exceeds the time horizon acceptable for private investors and commercial banks.
- Basic port infrastructure is immobile and has few alternative uses.

This set of characteristics is the main reason for financial involvement of governments in port construction and expansion projects.

INTERACTION WITH PORT CITIES

Ports and the cities of which they are a part interact across many dimensions: economic, social, environmental, and cultural. Any port reform process should take into account the linkages between city objectives and the port objectives. Transport integration the smooth transfer of cargo and equipment from land to water-borne systems is an essential port function, but it does not take place in isolation. A seaport node within a multimodal transport system is frequently associated with the development of an urban center and generates substantial employment, industrial activity, and national and regional development.

Many big cities trace their roots to the establishment of a port. This does not mean, however, that the port will be extended at the place where it was originally founded. Antwerp and Rotterdam are examples of ports that developed relatively close to the cities' central cores. Over time, however, they shifted operations away from city centers. The underlying reason was the increase in ship sizes (requiring deeper drafts and longer berths). Another reason contributing to the weakening of links between port and city centers is the rapid mechanization and specialization of port work and the accompanying increase in the operational scale and scope. These shifts led to increased storage space requirements and make ports very spaceintensive.

Another factor is the rapid industrialization of most developed country cities. The new industries emerging after World War II required large areas of land, preferably close to deep water, which often could not be found within the original port borders. Therefore, Maritime Industrial Development Areas (MIDAs) were located at some distance from old city centers. Technological changes and consequential port relocation have left substantial areas available for redevelopment for other purposes. Such areas are often located near city centers because that is where the port (and city) began. Therefore, land values are potentially high, although probably depressed prior to redevelopment because of the presence of decaying port facilities.

Three approaches commonly have been used for the development of surplus port land:

- Retaining it within the port authority for redevelopment as in the case of the Port of Barcelona. This implies a widening of the port's function from that of a port into a property developer. Such change may require modifications to the statutes of the public port authority, or of the trust port. The experience of Associated British Ports shows that when the port is in private hands, it is capable of effective development of surplus lands. The Port Authority of New York and New Jersey is an example of a public port authority with wide redevelopment powers.
- Transferring it to the local authority or municipality for redevelopment. In practice this is not always effective because the municipality might lack the resources to realize the full value of the land in question. On the other hand, there are examples (such as Baltimore and Rotterdam)

of successful regeneration by the municipality of port lands near the city center.

- Creating a special development corporation for the specific purpose of redeveloping an old dock area. This is most appropriate when the area is very extensive, involves various municipalities, and involves high redevelopment costs. An example of a separate corporation established for this purpose is the Puerto Madera Corporation in Argentina, which is a joint venture by the City of Buenos Aires and the national government for the redevelopment of old city docks for mixed commercial, residential, and recreational use. Probably the biggest and best-known special purpose corporation (SPC) is the London Docklands Development Corporation (LDDC), created to redevelop the old docks of the Port of London. The LDDC was established by the government and endowed with extensive planning powers as a result of the inability of six riparian municipalities to agree on a coherent and feasible plan for the docks' redevelopment.

Finally, the interests of ports extend beyond local traffic and transport. Hinterland connections, nationally and internationally, rely on road, rail, and waterway links.

Both the port authority and the port city should use their influence to establish needed intermodal infrastructure and agreements. In addition, the port authority and the port city should collaborate to efficiently accommodate traffic flows and limit transport costs (including external costs).

ROLE OF A PORT AUTHORITY

Ports usually have a governing body referred to as the port authority, port management, or port administration. Port authority is used widely to indicate any of these three terms. The term port authority has been defined in various ways. In 1977, a commission of the European Union (EU) defined a port authority as a State, Municipal, public, or private body, which is largely responsible for the tasks of construction, administration and sometimes the operation of port facilities and, in certain circumstances, for security. This definition is sufficiently broad to accommodate the various port management models existing within the EU and elsewhere. Ports authorities may be established at all levels of government: national, regional, provincial, or local. The most common form is a local port authority, an authority administering only one port area. However, national port authorities still exist in various countries such as Tanzania, Sri Lanka, Nigeria, and Aruba.

The United Nations Conference on Trade and Development (UNCTAD) Handbook for Port Planners in Developing Countries lists the statutory powers of a national port authority as follows (on the assumption that operational decisions will be taken locally):

- *Investment:* Power to approve proposals for port investments in amounts above a certain figure. The criterion for approval would be that the proposal was broadly in accordance with a national plan, which the authority would maintain.
- *Financial policy:* Power to set common financial objectives for ports (for example, required return on investment defined on a common basis), with a common policy on what infrastructure will be funded centrally versus locally, and advising the government on loan applications.
- *Tariff policy:* Power to regulate rates and charges as required to protect the public interest.
- *Labor policy:* Power to set common recruitment standards, a common wage structure, and common qualifications for promotion; and the power to approve common labor union procedures.
- *Licensing:* When appropriate, power to establish principles for licensing of port employees or agents.
- *Information and research:* Power to collect, collate, analyze, and disseminate statistical information on port activity for general use, and to sponsor research into port matters as required.
- *Legal:* Power to act as legal advisor to local port authorities.

Increasingly, central governments implement seaport policies through the allocation of resources rather than through the exercise of wide-ranging regulatory powers.

While central governments should pursue macroeconomic objectives through an active seaport policy, port authority objectives should be more narrowly focused on port finances and operations. It is a widely accepted opinion among port specialists that a port authority should have as a principal objective the full recovery of all port-related costs, including capital costs, plus an adequate return on capital. *The full recovery of costs will help a port authority to:*

- Maintain internal cost discipline.
- Attract outside investment and establish secure long-term cash flows.
- Stimulate innovation in the various functional areas to guarantee a long-term balance between costs and revenues, especially when faced with innovations by terminal operators, port users, rival ports, and hinterland operators.
- Generate internal cash flows needed to replace and expand port infrastructure and superstructure.
- Compete according to the rules of the market system, without excessive distortions of competition.
- Put limits on cross-subsidization, which may be rational from a marketing point of view (market penetration, traffic attraction), but which can undermine financial performance.

- Avoid dissipation of the port authority's asset base to satisfy objectives of third parties (for example, port users demanding the use of land in the port area without regard to the land's most economic use or port and city administrations using port authority assets to pursue general city goals).

Full cost recovery should be viewed as a minimum port authority objective; once this objective has been achieved, however, the port authority can pursue other-than-financial objectives considered desirable by the government or by itself.

ROLE OF PORT OPERATORS

Just as central governments and port authorities play key roles in the port communities, so too do private port operators (such as stevedoring firms, cargo handling companies, and terminal operators). Port operators typically pursue conventional microeconomic objectives, such as profit maximization, growth, and additional market share. Only if port operators are free to pursue such objectives can the benefits of a market-oriented system be achieved.

Roles of a Transport Ministry

In a market-oriented economic system, the ministry of transport typically performs a variety of functions at a national level.

With respect to coastline and port issues, the main tasks and responsibilities of the ministry can be summarized as follows:

- *Policy making:* The ministry develops transport and port policies related to:
- ~ Planning and development of a basic maritime infrastructure, including coastline defences (shore protection), port entrances, lighthouses and aids to navigation, and navigable sea routes and canals.
- ~ Planning and development of existing and new port areas (location, function, or type of management).
- ~ Planning and development of port hinterland connections (roads, railways, territorial waterways, and pipelines).
- *Legislation:* The ministry drafts and implements transport and port laws, national regulations, and decrees.
- It is responsible for incorporating relevant elements of international conventions (for example, the International Convention of Safety for Life at Sea [SOLAS], United Nations Convention on the Law of the Sea, the International Convention for the Prevention of Pollution from Ships [MARPOL]) into national legislation for signature members.
- *International relations:* Specialized departments of the ministry represent the country in bilateral and multilateral port and shipping forums. The ministry may also negotiate agreements with neighboring countries relating to water-borne or intermodal transit privileges.

- *Financial and economic affairs:* A ministerial department is usually responsible for planning and financing national projects. In many countries, a ministry of transport also finances basic port infrastructure as well as roads, waterways, and railways connecting ports with their hinterland.
- It should be able to carry out financial and economic analyses and assess the socioeconomic and financial feasibility of projects in the context of national policies and priorities.
- *Auditing:* These functions should be performed independently from the affected line organization and are usually included in a staff office. The auditors should report directly to the minister.

In many countries, transport directorates are established as independent bodies within a ministry and perform an executive function. They are usually responsible for one of the modes of transport, for example, the maritime and ports directorate (maritime administration).

The principal elements of a typical maritime and ports directorate are:

- Ship inspections and register of shipping (oversight of ship safety and manning conditions).
- Traffic safety and environment (safe movement of shipping and protection of the marine environment).
- Maritime education and training (maritime academies, merchant officers exams, and licensing of seafarers).
- Ports (execution of national port policy).
- Hydrotechnical construction (construction of protective works, sea locks, port entrances, and others).
- Port state control on the basis of the Paris and Tokyo memorandum of understanding terms.
- Security compliance (International Ship and Port Facility Security Code).
- Investigation into and adjudication of any maritime incident, such as fire on board a vessel, collision, stranding, piracy, or similar event.
- Performance of regulatory and licensing functions in respect to structures, partly or entirely founded on the seabed within the territorial waters, in the exclusive economic zone of a country, or in any navigable water or on any beach within the territory of a country.
- Vessel traffic systems and aids to navigation (construction and maintenance).

PORT FUNCTIONS

Within the port system, one or more organizations fill the following roles:

- Landlord for private entities offering a variety of services.
- Regulator of economic activity and operations.

- Regulator of marine safety, security, and environmental control.
- Planning for future operations and capital investments.
- Operator of nautical services and facilities.
- Marketer and promoter of port services and economic development.
- Cargo handler and storer.
- Provider of ancillary activities.

In view of the strategic significance of land, port property is rarely sold outright to private parties because of its direct and indirect effects on regional and often national economy and public welfare, its intrinsic value, and possible scarcity. Therefore, a key role for many port authorities is that of the landlord with the responsibility to manage the real estate within the port area. This management includes the economic exploitation, the long-term development, and the upkeep of basic port infrastructure, such as fairways, berths, access roads, and tunnels.

Port authorities often have broad regulatory powers relating to both shipping and port operations. The authority is responsible for applying conventions, laws, rules, and regulations. Generally, as a public organ it is responsible for observance of conventions and laws regarding public safety and security, environment, navigation, and health care. Port authorities also issue port bylaws, comprising many rules and regulations with respect to the behaviour of vessels in port, use of port areas, and other issues. Often, extensive police powers are also assigned to the port authority.

The planning function of the port authority in coordination with the municipality is a complicated affair, especially for large ports located within or near a city.

The port planner has to consider:

- The consistency of plans with the general terms of land use that have been set by the competent authority.
- The impact of port development proposals on the immediate surroundings (environment, traffic, facilities, and roads).
- The appropriateness of port development proposals in the context of international, national, and regional port competition.

Actual port services and balancing of supply and demand occur at the levels of the port authority and individual port firms. Hence, the development of realistic investment projects for infrastructure and superstructure should be initiated at these levels. Investment plans of industrial and commercial port operators or projects for specific cargo handling, storage, and distribution should be integrated at the level of the port authority to arrive at a strategic master plan for the port.

The individual master plans may then be integrated into a national seaport policy, taking into account macroeconomic considerations. Integration of individual master plans will help to avoid duplication of expensive, technologically advanced facilities when different ports in a national system strive to attract the same customers as well as ensure the selection of the appropriate locations for specific seaport facilities that will interconnect maritime and land transport systems.

To conclude, central governments should establish a national port policy that supports national economic objectives and creates a reasonable framework for port development. The development of plans for specific port projects, however, should remain in the hands of port operators. Oversight of nautical operations should be within a port authority's mandate and is often referred to as the harbormaster's function.

It generally comprises all legal and operational tasks related to the safety and efficiency of vessel management within the boundaries of the port area. The harbormaster's office allocates berths and coordinates all services necessary to berth and unberth a vessel. These services include pilotage, towage, mooring and unmooring, and vessel traffic services (VTS). Often, the harbormaster is also charged with a leading role in management of shipping and port-related crises (for example, collisions, explosions, natural disasters, or discharge of pollutants). In view of its general safety aspects, the harbormaster's function has a public character.

The cargo handling and storage function comprises all activities related to loading and discharging seagoing and inland vessels, including warehousing and intraport transport. A distinction typically is made between cargo handling on board of the vessel (stevedoring) and cargo handling on shore (landside or quay handling). Terminal operators can fulfill both roles. There are typically two types of cargo handling and terminal operating firms. The more common structure for terminal operating firms is a company that owns and maintains all superstructures at the terminal (for example, paving, offices, sheds, warehouses, and equipment). Other firms only use the superstructure or equipment that is owned by the port. Such firms typically only employ stevedores or dock workers and have virtually no physical assets.

The port marketing and promotion function is a logical extension of the port planning function. Port marketing is aimed at promoting the advantages of the entire port complex for both the port authority to attract new clients and for the port industry to generally promote its business. This type of broad marketing is distinct from customer-oriented marketing that is aimed at attracting specific clients and cargoes for specific terminals or services. A variety of ancillary functions such as pilotage, towage and ship chandlering, fire protection services, linesmen services, port information services, and liner and shipping agencies exist within the port community. Large port authorities usually do not provide these services, with the possible exception of pilotage and towage. In a number of smaller ports, however, these are part of the port authority operations because of the limited traffic base.

PORT ADMINISTRATION MODELS

A number of factors influence the way ports are organized, structured, and managed, including:

- The socioeconomic structure of a country (market economy, open borders).
- Historical developments (for example, former colonial structure).
- Location of the port (urban area or in isolated regions).
- Types of cargoes handled (liquid and dry bulk, general cargo, or containers).

Four main categories of ports have emerged over time, and they can be classified into four main models: the public service port, the tool port, the landlord port, and the fully privatized port or private service port.

These models are distinguished by how they differ with respect for such characteristics as:

- Public, private, or mixed provision of service.
- Local, regional, or global orientation.
- Ownership of infrastructure (including port land).
- Ownership of superstructure and equipment (particularly ship-to-shore handling equipment, sheds, and warehouses).
- Status of dock labor and management.

Service and tool ports mainly focus on the realization of public interests. Landlord ports have a mixed character and aim to strike a balance between public (port authority) and private (port industry) interests. Fully privatized ports focus on private (shareholder) interests.

SERVICE PORTS

Service ports have a predominantly public character. The number of service ports is declining. Many former service ports are in transition towards a landlord port structure, such as Colombo (Sri Lanka), Nhava Sheva (India), and Dar es Salaam (Tanzania).

However some ports in developing countries are still managed according to the service model. Under it, the port authority offers the complete range of services required for the functioning of the seaport system. The port owns, maintains, and operates every available asset (fixed and mobile), and cargo handling activities are executed by labor employed directly by the port authority. Service ports are usually controlled by (or even part of) the ministry of transport (or communications) and the chairman (or director general) is a civil servant appointed by, or directly reporting to, the minister concerned.

Among the main functions of a service port are cargo handling activities. In some developing country ports, the cargo handling activities are executed by a separate public entity, often referred to as the cargo handling company. Such public companies usually report to the same ministry as the port authority.

To have public entities with different and sometimes conflicting interests reporting to the same ministry, and forced to cooperate in the same operational environment, constitutes a serious management challenge. For this reason, the port authorities and cargo handling companies of Mombasa, Kenya, and Tema and Takoradi, Ghana, were merged into one single entity.

TOOL PORTS

In the tool port model, the port authority owns, develops, and maintains the port infrastructure as well as the superstructure, including cargo handling equipment such as quay cranes and forklift trucks. Port authority staff usually operates all equipment owned by the port authority. Other cargo handling on board vessels as well as on the apron and on the quay is usually carried out by private cargo handling firms contracted by the shipping agents or other principals licensed by the port authority. The Port of Chittagong (Bangladesh) is a typical example of the tool port.

The Ports Autonomes in France are also examples, in particular the container terminals, which are managed and operated along the principles of the tool port, although for more recent terminals the private terminal operators have made the investment in gantry cranes. This arrangement has generated conflicts between port authority staff and terminal operators, which has impeded operational efficiency.

The above-mentioned division of tasks within the tool port system clearly identifies the essential problem with this type of port management model: split operational responsibilities. Whereas the port authority owns and operates the cargo handling equipment, the private cargo handling firm usually signs the cargo handling contract with the shipowner or cargo owner. The cargo handling firm however, is not able to fully control the cargo handling operations itself. To prevent conflicts between cargo handling firms, some port authorities allow operators to use their own equipment (at which point it is no longer a true tool port). The tool port has a number of similarities to the service port, both in terms of its public orientation and the way the port is financed.

Under a tool port model, the port authority makes land and superstructures available to cargo handling companies. In the past, these companies tended to be small, with few capital assets. Their costs were almost entirely variable. The cost of underuse of port facilities was usually absorbed by the port authority, which minimized risk for the cargo handling companies. Often, the provision of cargo handling services was atomized, companies were small with activity fragmented over many participants. The lack of capitalization of the cargo handling companies constituted a significant obstacle to the development of strong companies that could function efficiently in the port and be able to compete internationally. However, with the above in mind, a tool port does have its advantages, particularly when it is used as a means of transition to a landlord port. Using the tool port model as a catalyst for transition can be an attractive option in cases where the confidence of the private sector is not fully established and the investment risk is considered high. A tool port may mitigate this by reducing initial capital investment requirements. Another example could include a government looking to expedite port reform initiatives, but requires extensive amounts of time for legal statutes to be established. Laws and regulations for establishing a tool port may be less extensive since no state assets are being transferred to the private sector, and therefore make it an easier model to adopt in the first phase of reform.

LANDLORD PORTS

The landlord port is characterized by its mixed public-private orientation. Under this model, the port authority acts as regulatory body and as landlord, while port operations (especially cargo handling) are carried out by private companies. Examples of landlord ports are Rotterdam, Antwerp, New York, and since 1997, Singapore. Today, the landlord port is the dominant port model in larger and mediumsized ports.

In the landlord port model, infrastructure is leased to private operating companies or to industries such as refineries, tank terminals, and chemical plants. The lease to be paid to the port authority is usually a fixed sum per square meter per year, typically indexed to some measure of inflation. The level of the lease amount is related to the initial preparation and construction costs (for example, land reclamation and quay wall construction). The private port operators provide and maintain their own superstructure including buildings (offices, sheds, warehouses, container freight stations, workshops). They also purchase and install their own equipment on the terminal grounds as required by their business. In landlord ports, dock labor is employed by private terminal operators, although in some ports part of the labor may be provided through a portwide labor pool system.

FULLY PRIVATIZED PORTS

Fully privatized ports (which often take the form of a private service port) are few in number, and can be found mainly in the United Kingdom (U.K.) and New Zealand. Full privatization is considered by many as an extreme form of port reform. It suggests that the state no longer has any meaningful involvement or public policy interest in the port sector. In fully privatized ports, port land is privately owned, unlike the situation in other port management models. This requires the transfer of ownership of such land from the public to the private sector. In addition, along with the sale of port land to private interests, some governments may simultaneously transfer the regulatory functions to private successor companies. In the absence of a port regulator in the U.K., for example, privatized ports are essentially self-regulating. The risk in this type of arrangement is that port land can be sold or resold for non-port activities, thereby making it impossible to reclaim for its original maritime use. Moreover, there is also the possibility of land speculation, especially when port land is in or near a major city. Furthermore, sale of land to private ports may also sometimes raise a national security issue.

The U.K. decided to move to full privatization for three main reasons:

- To modernize institutions and installations, both of which often dated back to the early years of the industrial revolution, to make them more responsive to the needs and wishes of the users.
- To achieve financial stability and financial targets, with an increasing proportion of the financing coming from private sources.
- To achieve labor stability and a degree of rationalization, followed by a greater degree of labor participation in the new port enterprises.

Box 5 summarizes the strong and weak points of the principal port management models. Box 6 outlines the sectors (public or private) and their various responsibilities under the four basic port management models.

GLOBALIZATION OF TERMINAL OPERATIONS

Port authorities are increasingly confronted with the globalization of terminal operations. During the 1990s, a number of terminal operators and major shipping lines merged to invest in and take control of a large number of terminals all over the world. This trend has far reaching consequences for the strategic position of port management in relation to some of their major clients. This trend towards globalization has affected mainly containerized operations. Today, a handful of major carrier alliances and independent terminal operators increasingly dominate the major global container trades. The global carriers have sought to secure their competitive positions by concluding long-term contracts for dedicated container terminals in major, strategically located ports. Their reasoning is that they believe they need to control all stages of the transport chain to remain competitive. These efforts to establish integrated transport chains pose a challenge for port authorities in their relations with the larger carriers. For example, how should a port respond if a large container operator demands to operate a dedicated terminal and threatens to leave the port when it does not get its way?

It should be emphasized that full control of the transport and logistics chain by one consortium (a global monopolist) is not a desirable development. Because of regulatory measures by the United States and the EU, the complexity of the transport and logistics chain, and the number of players, a carrier's ability to control of the full chain seems like an illusion. However, some alliances may attain a significant degree of market dominance. Box 7 lists the fleets of the major container carriers, showing the number of vessels operated, the capacity expressed in TEUs, and the number of vessels under construction. The container shipping market is still much commoditized compared to other industries (energy, rail, and the like) with global market shares of the largest carrier not exceeding 18-19 percent (2005). However, the carrier industry, as well as the terminal operator industry, is moving towards greater consolidation and larger global players and operators are emerging.

Competition between major carriers is intense. The scale of investment in a new generation of container vessels represents a massive commitment. To fill these vessels, the carriers try to secure local control and coordination over inland cargo haulage and feeder operations. In this way, they try to secure their market share and meet perceived service needs. Port handling charges are considered as being of secondary importance in achieving these goals.

Relationships between ports and carriers fall into four broad categories:

- First are ports that face strong interport competition in the container handling sector. Container lines may easily shift operations to other ports if their financial and operational demands are not met. To attract major container lines, the port authority may offer them dedicated facilities while other, smaller lines are accommodated at common user terminals. Without such dedicated facilities, major lines could move to other competing ports. Examples of this category are the Ports of Yokohama and Long Beach.
- Second are ports that derive the bulk of their business from a major container line, and therefore, are dominated by this client. If the dominant line were to abandon the port, 80-90 percent of the traffic could be lost. Examples of such ports are Algeciras and Salalah.
- Third are ports where, although no single shipping line may dominate the port's traffic volume, there is a possibility for that line to pressure the port authority into accepting a dedicated terminal because of competition for transit traffic in the larger region. An example of this type of port is Miami, which is a hub for the Caribbean and Central and South America. Competitors include Kingston (Jamaica) and Freeport (The Bahamas). As the competitive positions of these ports improve, carriers may increase pressure on Miami to grant dedicated terminals.
- Fourth are major world ports such as Shanghai, Hong Kong, Singapore, and Rotterdam. Such ports have a very welldeveloped container sector. Initially, these ports resisted pressures from shipping lines to accept dedicated terminals. However in Rotterdam, the large Europe Container Terminal (ECT) has been acquired by Hutchison Port Holdings (HPH), which was obliged by the European Commission to sell ECT a 33 percent share in the Maersk Delta Terminal. Also at Maasvlakte (Rotterdam), P and O Nedlloyd started the construction of its Euromax Terminal, which is expected to be

operational in 2008. Thus the Port of Rotterdam currently accommodates a mix of dedicated and common user terminals. In Antwerp, developments are similar.

The Port of Singapore did not meet the requests of Maersk Line, which resulted in the carrier initiating the development of the nearby Malaysian Port of Tanjung Pelepas with its affiliate A. P. Moller Terminals, which conducts business under the name APM Terminals. However, in this particular case it should be noted that the container operations in Singapore are carried out by PSA Corporation, which itself is competing globally in the container terminal market.

The changes in terminal management are fast. Container lines may use a common user terminal with the advantage that they can switch easily to a competing facility when the need arises, which has competitive advantages. On the other hand, major container carriers are increasingly interested in securing berth and throughput capacity, with the larger ones aiming at operating their own dedicated terminals directly or through affiliated global terminal operators. Strategic alliances between global terminal operators and major container lines are likely to continue in the near future. With such consolidation and alliances increasing in the industry, there is the growing concern of dominant market shares or monopolies or oligopolies developing at both local and regional levels. Governments should be aware of these trends and the impacts.

Apart from the major container lines, a number of global terminal operators have also emerged during the 1990s and the top 10 have distanced themselves from the rest of the market over the last three to five years (see Box 8). These companies operate a large number of terminals all over the world. Their main objective is not to control the transport chain, but to make a profit by offering terminal services. However, when too many terminals within a region are controlled by one operator, the competent authority or government agency may decide that special regulatory measures are needed to protect against the danger of a monopoly. This was the case in Rotterdam when Hutchison Port Holdings (Hutchison-HPH) bought 49 percent of the shares of ECT. The European Commission decided to refuse permission for this transaction on the grounds that this would have allowed Hutchison to establish a dominant market position in Northwestern Europe since Hutchison already owned Felixstowe, Thamesport, and Harwich.

PORT MANAGEMENT AND COMPETITION

Competition within and between ports has a bearing on the management structure of the port and the relations between the port authority and the terminal operators and cargo handling companies. These changing relations are often cited as an important reason for changing the port management structure. Many port authorities consider the creation of competitive conditions among port operators the cornerstone of their port policy.

One can distinguish between interport competition (competition between different ports) and intraport competition (competition between different enterprises within one port complex).

To reduce the risk of monopolies, port authorities usually stimulate intraport competition. However, medium-sized and smaller ports, because of their limited traffic, often accommodate only one port terminal operator. In such cases, port authorities often use their quasi-governmental powers to regulate port charges and tariffs.

Key factors affecting interport competition include:

- *Geographic location:* A port that is strategically located close to wellestablished transport routes has competitive advantages. A strategic location typically possesses at least the following characteristics:

~ Proximity to one or more major maritime routes.

~ Natural deep water, good protection against waves and currents, large waterfront and landside expansion possibilities.

~ Proximity to major production or consumption areas.

~ Good hinterland connections (road, rail, pipeline, and waterway) with high frequency service offering good connectivity.

- *Legal framework:* The well-balanced national and local legal framework applicable to port management greatly bolsters investor confidence. Many countries have enacted specific port laws dealing with powers and responsibilities of the various actors in the sector. Moreover, land and competition laws are equally important, as well as an independent judiciary.
- *Financial resources:* A port with sufficient financial means of its own or the capacity to raise the funds required to develop and improve the port has a competitive advantage over ports with limited resources or no financial autonomy.
- *Institutional structure and socioeconomic climate:* The management structure of the port must be conducive to private sector investment. Related to this is the socioeconomic climate in the port; private investors prefer ports with a sufficient and well-trained labor force and good relations between employees and employers.
- *Efficiency and price:* Various investigations indicate that port costs are an important, although not decisive, factor in making choices, especially for cargo owners or their representatives. In a world where manufacturers seek to trim costs and improve customer service through the adoption of sophisticated logistics processes, efficiency and the price-performance ratio are increasingly important.

- *Image of the port:* The image the port projects is another factor in its competitiveness. The preferred image is an optimum mix of the above-mentioned components.

PORT SECTOR REGULATOR

When interport competition is muted or absent, port authorities or public or private terminal owners are apt to use their monopoly market positions to raise tariffs (in particular for captive cargoes), which may justify regulation. The need for such regulation may lead to the creation of an independent port sector regulator. The objectives of the port sector regulator are to ensure fair competition among competing operators in the port; to control monopolies (including public ones) and mergers; and to prevent anticompetitive practices.

A port sector regulator typically has legal powers to counter anticompetitive practices, such as:

- Use of a dominant position to prevent or lessen competition.
- Cross-subsidization by the provider of monopoly services of contestable services, thereby threatening fair competition.
- Price fixing among competitors.
- Use of other practices that are intended to restrict, distort, or prevent competition.

Smaller ports are more vulnerable to anticompetitive abuses because their traffic volumes limit the number of container, bulk, and oil terminals. Generally, when a monopoly or merger situation does not operate against the public interest, it may be permitted provided it is properly regulated. Examples of regulation in such cases could include tariff caps, volume or traffic thresholds to trigger any additional future concession, or expansion limits to incumbent operators that otherwise require an open tender.

The establishment of a port sector regulator should only be effected in the event of serious threats to free competition within the port. It should preferably have the character of an arbitrator instead of a court of law, and be accepted by the port community as being independent. For a more detailed discussion of the economic regulation of ports, see Module 6.

VALUE-ADDED SERVICES

Generally, the function of a port as a node in the transport chain depends on its location and on the economic and technical developments that exist in its hinterland. Modern production techniques and consumption patterns increase the use of transportation systems beyond levels suggested purely by the growth in trade and commerce. As a result, more specialized handling, storage, and other logistics facilities are needed. More and more, ports are becoming part of integrated logistics chains. This process of specialization and changing demands, which has taken place over the last two decades in most Western countries, is now taking place with even greater speed in new market economies.

From the port's point of view, creating new services boosts economic performance as well as its attractiveness to existing and potential clients. This, in turn, can help maintain and improve a port's competitive position. When assessing the wisdom of developing new services, it is important to pay attention to the valueadding potential of the services. This potential can vary product by product and activity by activity. Numerous activities can be classified as value-added services (VAS). Box 11 identifies a number of them. VAS can be divided into value-added logistics (VAL) and value-added facilities (VAF). VAL has two major components: general logistics services (GLS) and logistics chain integration services (LCIS). GLS are, among other activities, loading and unloading, stuffing and stripping, storage, warehousing, and distribution. These are the more traditional logistics activities and do not directly affect the nature of the product as it moves through the port.

Beyond these traditional activities, more complex LCIS are being developed. To carry out activities that manufacturers do not consider part of their core business, logistics service providers may take over parts of the production chain (for example, assembly, quality control, customizing, and packing) and after sales services (for example, repair and reuse). However, LCIS are only appropriate for certain types of goods. The products that have the highest potential to benefit from such services include consumer electronics, pharmaceutics, chemical products (except for those carried in bulk), clothing, cosmetics and personal care products, food, machinery, and control engineering products.

The second group of VAS, that is, VAF, is very diverse. These types of activities cannot generally be assigned to a particular type of product or freight flow. It is possible, however, to impute a certain VAF potential by analyzing freight flows such as dry and liquid bulk, general cargo, containerized cargo, and roll-on roll-off. A large container throughput might create the economic basis for establishing container repair facilities, handling vast quantities of chemicals requires port reception facilities, and substantial roll-on roll-off traffic might justify truck maintenance and repair shops. Box 12 broadly depicts the potential for both VAL and VAF activities for different types of cargoes.

Containerized and general cargoes typically have the highest VAL potential. GLS and LCIS have the best opportunity to serve these cargoes. The VAL potential for roll-on roll-off is very limited. Trucks with drivers are too expensive to be delayed while the cargo is modified; additionally, these loads are usually customer tailored. VAF, such as tanking, cleaning, repair, parking, security, renting, and leasing facilities have a better potential to serve the roll-on rolloff market. Dry and liquid bulk flows have the lowest potential for both VAL and VAF.

To provide a favourable environment for VAL and VAF, many ports are developing distriparks. A distripark is an area where companies are established to perform trade and transportrelated value-added services and can also include locations within the port's larger hinterland region. There is no standard development plan for a distripark. As can be seen from the various developments in the Netherlands, France, Germany, and the U.K. for instance, there is a large variety in distriparks. For example, in Rotterdam, there are three distriparks. The oldest one (Eemhaven) is devoted to container cargo distribution, the second one (Botlek) is devoted mainly to chemicals, and the third and most recent one is also dedicated to containerized cargoes, and includes large warehouses containing goods for European distribution (for example, Reebok).

PORT FINANCE OVERVIEW

Before 1980, service ports and tool ports were mainly financed by the government. The general infrastructure of landlord ports typically was financed jointly by the government and the port authority, and the terminal superstructure and equipment by private operators. Fully privatized ports were the exception. In the event a government had no funds for expensive port infrastructure, either port development was halted or money was acquired at preferential rates from an IFI such as the World Bank. Ports require expensive infrastructure to be able to compete successfully. Until recently, port authorities mainly relied on contributions and subsidies from national governments for building or improving basic port infrastructure. Such contributions usually were excluded from port financial accounts and therefore helped ports to exhibit positive financial positions.

Whether national governments finance basic port infrastructure depends on the government's political and economic policies. For example, if ports are considered part of the general transport infrastructure of the country, then investments in them may be considered to promote the national interest. Research shows that in 63 percent of the top container ports, the public sector (either the national government or the public port authority) was responsible for creating and maintaining (public) basic port infrastructure. In some countries, financing basic infrastructure is considered a public task (for example, in France, Italy, and Croatia) because this part of infrastructure belongs to the public domain, which is protected by law. To carry out construction activities or port operations in this domain, a public license is required. This requirement could reduce intraport competition if the licenses are granted only on a limited and discriminatory basis.

An often occurring problem with public (thus political) investment decisions is that the decision to invest does not necessarily originate at the same level of government as that of the financing sources and responsibilities. Because of this disconnect, the interest of public officials to increase efficiency and profitability of port assets is usually limited because they are not held accountable for the success or failure of their investment decisions. As mentioned earlier, the increasing role of private enterprise in the port sector exerts a direct influence both on port management and operations, as well as on the way capital projects are financed. The private sector has become interested in financing the construction of entire terminals, including quay walls, land reclamation, dredging, superstructure, and equipment. This has given rise to a large variety of financing and management schemes such as BOT (buildoperate- transfer), BOOT (build-own-operatetransfer), and BOO (built-own-operate). Each is designed to mobilize private capital while balancing public and private interests.

Government's views on ports are evolving. Increasingly, ports are considered separate economic entities, although still subject to national regional and local planning goals. As such, they should operate on a commercial basis. By the same token, subsidies for operational port infrastructure construction, such as port land, quay walls, common areas, and inner channels, should be avoided.

There still is, however, a category of port infrastructure for which it will be hard to find private investors: investments for expensive and long-lived infrastructure (for example, breakwaters and locks, entrance channels and fairways, and coastal protection works). The main stumbling block for private financing of such projects is their life span, which often exceeds 100 years, and the sunk investment aspect of these projects. Cost recovery of such works often cannot be achieved in 20 to 30 years (see Module 4), which is a normal repayment period for long-term loans for infrastructure works by IFIs. Nevertheless, the second- and third-order benefits from such infrastructure investments for national and regional economies may be substantial. Hence, many governments are still willing to finance part or all of long-term port investments as these contribute to the achievement of public policy objectives. Caution is warranted, however, whenever governments contemplate underwriting such investments.

FINANCING PORT PROJECTS

To further clarify financing approaches, it is important to distinguish among investments in basic port infrastructure, operational port infrastructure, port superstructure, and port equipment. Understanding these distinctions will help in deciding which investments should be paid for by the port and which should be paid for by the local or regional community, the central government, and private investors. Box 14 lists various types of port assets under these four categories.

In addition to financing the construction, rehabilitation, acquisition, and maintenance of physical assets, ports may also need to finance organizational restructuring and associated labor compensation as well as working capital to support operations. Each of these categories and their potential sources of financing are discussed below. In many countries, the government is responsible for financing basic infrastructure, either directly or through a contribution to offset its cost when the project is conducted, for example, by a highway authority or a port authority. In the Netherlands, construction of maritime access and protection works used to be carried out by and for the account of the government with the port authorities obliged to pay onethird of the relevant costs. In France, this issue is regulated in the Port Authority Law of 1965 (Law No. 65-491 of June 29, 1965), which allocates a minimum of 80 percent of the costs of basic port infrastructure of the Autonomous Ports to the national government. For the government, there are two key issues associated with making large direct investments in port facilities: how to find the necessary funds and how to recover the investment.

The ways in which the government (or any other public body) funds investments are diverse:

- Direct investments coming from the government investment budget.
- Direct investments coming from a special (port) fund.
- Loans from IFIs.

Direct investments, paid for by the investment budget or a special fund, are based on the assumption that they will have a substantial positive effect on the economy, as shown by the positive results of a cost-benefit analysis (always heavily dependent on traffic forecasts). For investments broadly benefiting the entire nation, it is not unusual that a government would not seek direct financial repayment.

However, there are also situations where the government may receive direct reimbursement for the funds it invested via a variety of rates and charges assessed against the beneficiaries of the investments.

These may take the form of:

- Compensation paid by the port authority in proportion to the volume of goods transported through a newly dredged channel (per ton or per TEU).
- A fixed amount per year paid by the port authority to the government.
- A percentage of the annual port dues paid by the port authority to the government.

Often, basic infrastructure elements are financed by an IFI under a government guarantee. However, even when IFI financing is made available, ports and governments must still face the challenge of providing matching shares for a period of 30 to 50 years and making interest payments over a period of some 20 years.

When considering financing of operational infrastructure, port authorities have a number of options from which to choose. For service ports or tool ports, governments will usually finance the operational infrastructure, with or without the assistance of an IFI. For landlord ports made up of self-contained terminals, investment in the terminal should be financed by the terminal concessionaire or the lessee, while the port provides the land (often in a condition ready for construction). The port may also provide the quay wall with the land, but, increasingly, private concessionaires have been willing to invest in this infrastructure.

Other financial arrangements are also common. For example, in U.S. public ports, the port authority may have access to 'cheaper' money than a private sector operator. In this case, the authority has the option to issue tax-free port revenue and general obligation bonds. Both give ports access to capital markets; the former relies on the revenues generated by operation of the new facility to repay debt, the latter assures purchasers of the debt that the government will make good on any repayments should revenues from operation of the new facility prove inadequate.

The most attractive situation, both from the point of view of the landlord port authority as well as of the operator, is the conclusion of a long-term lease contract with the operator (running for a period of 20 to 30 years) for the use of part of the port area.

This type of long-term lease has the legal character of a property right and has four advantages:

- At the end of the contract, possession of the land reverts to the government or port authority.
- The contract represents a property right that under certain conditions can be transferred to a third party. There usually is a clause in such contracts stating that such transfer of property rights requires prior permission from the port authority.
- All superstructures (buildings and equipment) may be financed and owned by the operator.
- It can be used as security for a bank loan.

For the financing of common areas (all areas within the port area not being part of a terminal or other port enterprise), the port authority may make use of retained earnings, issue its own bonds (where permitted to do so by its statutes and legal system) or make use of bonds, or simply take a bank loan. Except in the first case, the associated risk is with the borrower. The problem confronting public ports is what to use as collateral or guarantees for the lender, particularly since there may be restrictions with respect to the use of the port's assets.

In the event of a major reorganization programme for the port authority, substantial amounts of money may be required for compensation payments to personnel. (See Module 7 for a detailed discussion of labor issues affecting port reform.) Such payments often have a short payback period. Nevertheless, traditional sources of finance may be unwilling to lend money specifically for this purpose. There is, however, a possibility for triangular financing, that is, lending the money for some other transaction on condition that the funds thus liberated are used to compensate displaced workers. Moreover, a national government might be willing to provide funds for labor redundancy schemes with or without the involvement of an IFI.

Port operators and providers of services who take over existing installations and equipment from a port authority may have a greater need for working capital than investment capital, especially in their start-up periods. With respect to debt financing, operators face the problem of providing security because installations and equipment often may be leased under conditions that prevent them from being mortgaged. Since port operators are essentially private companies, an attractive alternative to debt financing is through the flotation of equity shares, the success of which will depend largely on the degree of confidence prospective shareholders have in the newly founded company and in its management.

Supplier credit, provided that it includes the financing of necessary spare parts over a period of at least three years, offers another potential source of funding for the procurement of equipment, with the usual limitations of this type of financing.

Finally, a joint venture between the port authority and the operator offers what may be an attractive source of finance for the operator. For a specialized terminal, where the likelihood of a competing terminal being constructed is remote, a joint venture may be reasonable. In most circumstances, however, the likely effect of a joint venture between a port authority and an operator is to obscure the transparency of the relationship between the different port functions and, more pragmatically, to discourage the entry of new operators to the port. Box 15 describes the challenges mounted by such relationships in the case of the Sri Lanka Port Authority.

FINANCING PORTS: FROM A LENDER'S POINT OF VIEW

Port authorities or port operators seeking to finance new facilities or equipment typically have to offer some sort of security to a prospective lender. Generally, they have assets and other support from political and business circles for the project they want to undertake. In many ports, however, land is government-owned and cannot be used to secure financing. And, when a port needs money to dredge a channel entrance to remain attractive and competitive, the channel itself does not constitute credible security for the lender. There are however, various options for ports to provide lenders 'comfort.'

Prospective lenders will examine closely the position of the borrower, which might be a port authority or a port enterprise. In the vast majority of cases, the latter are structured as limited liability companies. In the case of loans to a public port authority, the state or municipality usually provides a guarantee. A port authority might also be corporatized with the state or the port city as main shareholders. In both cases, the lender will assess the financial strength of the port authority and the public bodies owning it. This is often sufficient to ensure financing of the venture without too much regard to the assets supporting it. In Anglo-Saxon jurisdictions, a borrower may create a 'floating charge' (similar to a mortgage) over all assets. This avoids the need to consider specific elements of the port assets as collateral.

A port's most valuable asset is its land; however, land's value as a security for financing varies significantly. Generally the land is owned by a public body or by the port authority itself. In landlord ports, the land is concessioned or leased to private operators, with the exception of common areas, which usually have a low commercial value. In the majority of cases, port land cannot be mortgaged under a concession agreement. Sometimes it is legally possible to mortgage superstructure on the terminal. Using the land itself as collateral is therefore complicated. The land must have inherent worth and a user should be able to exploit it. If a right to use the port area concerned does not accompany the mortgage on port land, its value is considerably diminished. Another problem might be that the national legislation grants only limited rights to a mortgage. Lastly, in the event of a public port authority, the lender might be confronted with political processes complicating its ability to exercise rights under a mortgage. This makes the security less valuable to a lender.

In most ports, the concession or lease to private operators is the principal security for lenders, provided that the conditions of the concession or lease allow transfer of the contractual rights to another party. In the case of a full-fledged concession (including a BOT scheme), the financier often desires to have the ability to arrange for the operation of the terminal itself if the operator defaults. In the case of a concession or land lease, a port authority is usually obliged to transfer the concession or lease to a third party, such as transfer to another port-related firm, when certain conditions are met. This might be a cargo handling firm or terminal operating company, or a port-based industry such as a refinery or a chemical plant. Conditions attaching to the transfer typically require the new firm to use the facilities in accordance with their initial assignment and to generate sufficient seagoing traffic.

A port complex comprises a large variety of other assets that might be mortgaged or used as collateral, such as warehouses, quay cranes, offices and other buildings, tugs, dredged channels, and others. Some of these assets might provide security to a lender, especially when the assets can be used in other ports (for example, cranes and tugs). Others, because they are immobile or have few alternative uses, constitute little or no security (for example, dredged channels). An important aspect of securing financing is the legal right of a port operator to own buildings on land leased from the port authority. Lenders are usually prepared to finance buildings and certain types of equipment in view of their intrinsic value. Port firms, and sometimes privatized or corporatized port authorities, typically take the legal structure of a joint stock or limited liability company. The equity of such enterprises does not constitute security in itself, but may help to attract investment funds. Rights of equity holders to repayment usually rank immediately behind the rights of a lender. When balance sheet financing is undertaken, a high level of equity (in relation to debt) means that more funds are available to absorb losses before lenders come under threat.

One of the most important elements of financial security is the cash flow generated by the port or terminal. A lender almost always wants the earnings of the project to provide security for the loan. Estimation of such earnings is highly complex because it involves assessing elements such as future traffic levels, port revenues and expenses, the expected general economic development of the country, potential exchange rate risks, the future political climate, and other factors. The more accurate and reliable the traffic and financial forecasts are perceived to be by prospective investors, the higher the probability that a port authority or port operator will be able to attract risk capital and obtain loans.

Governments may also guarantee commercial loans against political risk and possibly use the guarantee programmes offered by the IFIs. In the port sector, lenders often take security via assignment of port charges. However, much will depend on the terms of the concession or lease agreement, terms of earlier financing, and the rights of third parties. Finally, financing can be affected by the provision of additional government support. A government may invest equity in a firm it deems essential for the general development of the port. It may also provide subordinated loans. Direct financial involvement of governments and public port authorities is increasingly common, despite potential conflicts of interest. Sometimes a government may assign certain rights or grant concessions such as a duty-free status (as was the case at Jebel Ali) to enhance the success of the venture. Properly focused government support can be very important to provide additional comfort to lenders.

PUBLIC-PRIVATE PARTNERSHIPS

As private sector involvement in financing port and other infrastructure works has increased, the tools for financing these facilities have become increasingly sophisticated and the legal conditions to be satisfied by the project more strict.

The private sector evaluates its participation in port infrastructure and superstructure projects based on the following elements:

- Expected yield.
- Adequate debt/equity financing structure (for example, 65/35, 70/30, 75/25).
- Strong sponsorship.

- Solid legal contracts.
- Transparent legal framework.
- Fair and open bidding procedures.
- Credible feasibility analyses (technical, institutional, financial, economic, and environmental).

Funding large infrastructure investments in greenfield port projects is more risky because of certain complicating factors, including:

- The large proportion of necessary equity contributions (for example, a minimum proportion of 60 percent) due to the high risk associated with long construction and payback periods.
- The difficulty of projecting future traffic volumes.
- The capital-intensive nature of the investments.
- The continuing risks associated with operations, such as a refusal of requests for tariff adjustments, changes in tax policy, or introduction of new handling techniques that make existing facilities obsolete.

PORT REFORM MODALITIES

Today, the term port reform connotes the changing institutional structure of the port business and the much greater involvement of the private sector in the exploitation and financing of port facilities, terminals, and services. Port reform, therefore, results in changing relationships between the public and private sectors. The sharp increase in world trade over the last 60 years focused the attention of national governments on the economic importance of ports. This was especially the case in major ports developing large industrial sites within their domain. In the 1950s and 1960s, many nations introduced institutional changes with the aim of coordinating port development at national and regional levels and preventing overinvestment in expensive port infrastructure. For example, the United Kingdom established its National Ports Council for this purpose.

In the former Soviet Union, Eastern Europe, and in many socialist-oriented developing countries the situation was entirely different. Ports were considered part of the national state structure (for example, as an element of the ministry of merchant marine or ministry of transport) and were often controlled by national shipping companies.

Every matter involving maritime policy was decided centrally, with port authorities carrying out the various day-to-day nautical and operating functions. At the beginning of the 1980s, the belief in the management and operating capacities of national governments faded in most market economy countries. Central structures came under fire and often lost some of their powers. The privatization wave launched in the late 1970s and early 1980s by Margaret Thatcher in the U.K. also affected the port sector and resulted in a reassessment of the role of the government and private enterprise.

The demise of the communist system in the beginning of the 1990s resulted in the virtual collapse of centrally controlled port systems in the former socialist countries. They too embarked on port reform and adapted the institutional and financial structure of their port sectors to market conditions.

Despite the social and economic reforms of the past 35 years, the public sector has retained a strong role in port development. Generally, in a market-oriented economy a government continues to be responsible for the development of public goods, goods that have a social utility, but that cannot be provided by the private sector because of low profitability. Moreover, another reason for continuing government involvement in the port sector is the strong ties to government responsibilities in the areas of land use planning, environmental protection, job creation, and the economic stimulation of underdeveloped areas.

Box 16 is a compilation of a considerable number of surveys seeking to summarize the most frequently cited reasons for change in the management or ownership of ports.

STRATEGIES AND REFORM OPTIONS

Many port managers and government officials believe that the only way to improve the performance of public port organizations is through the process of privatization. They hold this view because they believe that certain characteristics of the private sector are indispensable to achieve commercial success. The term privatization has therefore become synonymous (and confusingly so) with port reform. Privatization, however, more accurately refers to one aspect of port reform the introduction of the private sector into areas previously reserved to the public sector, finally resulting in the transfer of port land into full private ownership.

Governments and port managers can select from among a variety of strategies for improving organizational and operational performance, including:

- Modernization of Port Administration and management.
- Liberalization or deregulation port services.
- Commercialization.
- Corporatization.
- Privatization.

Each of these options may be equally valid and successful forms of port reform, depending on the setting of the port in question. Each of these options is defined below.

Modernization of port administration assumes that performance can be improved by introducing more suitable systems, working practices, or equipment and tools within the existing system of bureaucratic constraints. The advantage of this strategy is that certain changes in the organization can be made without the requirement to change laws or national policy.

Liberalization and deregulation are the reform or partial elimination of governmental rules and regulations that enable private companies to operate in an area where previously only the public sector was allowed to operate. In the case of commercialization, although the public port is not transformed into a private company, it is given more autonomy and made accountable for its decisions and overall performance. A commercialized port authority applies the same management and accounting principles as private firms and can adopt private sector characteristics and practices to become more customer oriented as well as more efficient and profitable. In the case of corporatization, a public port enterprise is given the legal status of a private company, although the public sector or government still retains ownership. All assets are transferred to this private company, including land lease rights. Land ownership usually remains with the port authority. The most complex form of reform is privatization. A useful definition of this term can be found in the UNCTAD publication of 1998 Guidelines for Port Authorities and Governments on the Privatization of Port Facilities: 'Privatization is the transfer of ownership of assets from the public to the private sector or the application of private capital to fund investments in port facilities, equipment, and systems.'

More specifically related to the port sector are two more variations of privatization:

- Comprehensive privatization: A scheme in which a successor company becomes the owner of all land and water areas as well as of all the assets within the port's domain (this is equivalent to the sale of an entire port to a private company).
- Partial privatization: A scheme in which only part of the assets and activities of a public port body are transferred to the private sector (such as the sale of existing berths, the transfer of pilotage or towage functions, or a concession by a public port authority to a private company to build and operate a terminal or a specialized port facility).

Hence, privatization expands the role of the private sector in the ownership or operations of existing port facilities and services, as well as in the development of new port facilities. In the following sections, the various port reform options are described in greater detail.

MODERNIZATION OF PORT ADMINISTRATION

The strategies of liberalization, commercialization, corporatization, and privatization all attempt to improve the efficiency of the port administration and the operations through the introduction of a business-like environment. Although these strategies can be effective, some governments are reluctant to implement them because they fear that such institutional modifications may lead to a disruption of services or loss of government authority, prerogatives, and power. As a result, governments sometimes prefer other less sweeping methods to improve organizational performance, such as the modernization of the port's administration. Such a strategy assumes that the performance can be improved even in the prevailing environment of bureaucratic constraints. The advantage of this strategy is that certain changes in the organization can be made without the necessity to make legal or policy changes.

Examples of improvements that can be introduced without legal or policy changes are:

- Adoption of corporate planning practices.
- Application of human resources development (HRD) planning.
- Use of computer applications and management information systems (MIS).
- Development of electronic data interchange (EDI) and information and communication technology.

Many ports have refrained from introducing corporate planning (strategic management or strategic planning) because port managers fear that its positive effects may be undermined by bureaucratic or cultural considerations.

Effective corporate planning is dependent on strategy formulation involving group interaction. While group-based strategic decisions often can offer the best available alternatives, a strict hierarchical organizational structure places the majority of important decisions in the hands of a single executive. In such cases, the success or failure of port development and policy is dependent on one person only, which is a risky situation. But this is precisely the most frequently observed form of management in traditional ports.

Career planning and management development are important elements in a port modernization strategy. Many ports have failed to introduce career planning and career development in the organization, or omitted to link the two activities. As a result, such organizations are characterized by low employee motivation levels, high absenteeism, and high turnover rates at management level positions. Efforts to improve the administrative environment and performance should include the rational use of computer applications and the application of modern communication technologies. Such developments are perhaps the most significant technological efforts undertaken by ports. Many have developed advanced computerized management information systems. EDI and information and communications technology are excellent tools to improve port administration and communication. In the final analysis, the modernization option generally has not led to fundamental changes in the port sector, which is what the reform process sets out to do. It should, therefore, be considered as a stepping stone towards a more comprehensive reform programme.

LIBERALIZATION

Liberalization sets the stage for a private organization to carry out certain port activities previously reserved exclusively for the public sector (public monopoly). With this reform, the private sector is authorized to provide selected port services to users in a competitive environment with the intent of increasing efficiency and improving port-client responsiveness. The essential feature of the liberalization option is implementing legislation that permits the private sector to provide facilities and services and to compete with the existing public port organization. The most important advantage of this system compared to other port reform systems is that the public port operator, even if inefficient, will continue to exist as a form of insurance against disruptions in service, while unsuccessful private port operators can be replaced.

Since liberalization may temporarily introduce competition between public and private port operators, the two must be able to compete effectively and fairly. This might require the introduction of an independent port sector regulator. Actually, the logic of liberalization should lead the public port authority to fully withdraw from commercial activities and concentrate on any necessary regulatory functions.

Liberalization is often opposed because of the existence of internal as well as external crosssubsidies. This, for instance, occurs when ports with a statutory monopoly cross-subsidize unprofitable services in competitive markets with profits earned in monopoly markets. For example, in many ports the most profitable activity is the container terminal operation, the revenues of which frequently support bulk or general cargo facilities and services. Other forms of cross-subsidy occur when a public port organization realizes substantial revenues from non-maritime-related activities, such as real estate development, and uses these revenues to underwrite port-related costs. With this type of support to draw on, the public organization has a competitive advantage over its private counterpart.

On the other hand, the price advantage that the public port body may have had diminishes as competition erodes its monopoly power and prices are set in a more competitive environment. Its price levels cannot match those of the private sector if it has to rely on inflated prices to subsidize other port services. The former monopoly may, as a consequence, be forced to scale back or cease the unprofitable activities (which, although unprofitable, may be vital to the nation) to compete effectively with the private sector.

On many occasions, the public sector continues to rely on public subsidies, thereby undermining fair competition between the public and the private sectors. This strongly argues for the clear separation of the regulatory and commercial roles in a port, with the port authority taking on the former and the private operator the latter.

Another potential problem associated with the liberalization option is the possibility that the public port organization will use other unfair practices to compete against private operators. The port authority, for example, may take actions that are beneficial to the public terminals, but are disadvantageous to the private terminals. One example is the dredging of certain Asian ports; often, the government ministry or the public port authority provides exclusive dredging services. This public entity can refuse to offer this service to the private operators, thereby putting those operators at a competitive disadvantage. Another possibility is that the service would be provided to the private sector at a higher price than the one charged to the public sector. To avoid such potential conflicts of interest, the government may also decide to liberalize or privatize these essential complementary services to create a level playing field. Because of these situations, the logical conclusion for the liberalization option is for all commercial activities of the port to be ultimately transferred to the private sector.

COMMERCIALIZATION

Commercialization is the introduction of commercial principles and practices into the management and operation of a port authority or part thereof, requiring it to operate under market disciplines. The process can be achieved through negotiated performance contracts between the government, acting as the owner of the port, and the port management. The agreement specifies the port's objectives in terms of performance goals, service quality, and social obligations.

Commercialization is characterized by the following:

- Decentralization of the decision-making process.
- Relaxation of the hierarchy of the port organization, thereby allowing port management to exercise much greater control over:
- Budgeting.
- Procurement and purchasing.
- Maintenance strategies and programming.
- Salary scales and employment conditions of labor and staff.
- Hiring and firing.
- Setting objectives and performance targets.
- Formulation of strategies.

Essentially, commercialization aims to create an environment in which the port authority runs on a commercial basis. This involves a variety of business-type decisions. The chief executive typically has a certain freedom of action and refers only specific matters relating to overall policy or strategy to the controlling body (the relevant ministry or city council). Commercialization is designed to allow port management to conduct, to a large extent, its own affairs and at the same time imposes on it responsibility and accountability for its decisions and performance. In practice, however, a common problem has been that governments continue to interfere in port decisions, undermining the authority of port management. Commercialization seeks to provide port managers with decision-making authority and responsibility similar to that existing in private sector organizations. However, since the port enterprise may still have substantial monopoly power, managers may not be confronted directly with the hardships and necessary discipline imposed by market competition. Therefore, a commercialized government organization often will not be as efficient as a comparable private firm, unless it is subject to competition.

Since the essence of commercialization is to require and empower port management to perform as well as the private sector, changes in the institutional and legal structures of the port organization are required to remove bureaucratic obstructions. A common first step in the process of commercialization and the elimination of bureaucratic inefficiencies is to transform the port organization into a truly autonomous port authority. Box 17 notes that the governments of China and Mexico followed this course.

Commercialization should result in the creation of a port authority board to oversee the organization's activities, removing that responsibility from the central government ministry or city. At the same time, however, the government may still need to exercise some form of oversight to safeguard the public interest.

Commercialized port authorities should:

- Be financially independent (own their assets, establish their own budgets, and make their own investment decisions).
- Have their own personnel schemes separate and distinct from the national civil service, patterned on the schemes of private companies.
- Have a management that is responsible for and held accountable for the port's performance by a board. Board members can be appointed by the national or local government, port users, or representative labor organizations.

In many countries, the process of commercialization is only partially implemented because procurement and contracting practices remain subject to national government regulations. A weakness of the commercialization process is that during its introduction, the acting public sector manager becomes the chief executive responsible for pushing through the changes in the organization. The manager's performance and commitment to the commercialization of the port authority greatly influence the management team and the shape and pace of reform.

In other words, managers accustomed to civil service procedures and practices have to drastically change their management styles. This has proven to be a difficult transition and is the reason why, in many such processes, managers with private sector experience soon replace the former civil service senior management. A wellthought- out training programme may be an effective tool to change attitudes and prepare management and staff for the different style and culture commercialization brings.

CORPORATIZATION OF TERMINALS

The next gradation on the path to full privatization is corporatization. Corporatization goes further than commercialization in that it involves the transformation of the public port authority or part thereof into a corporation. This means that the port authority or one or more of its constituent parts, such as a port authority-operated container or general cargo terminal, is converted into a legally and financially independent legal entity with its own board of directors.

The government or public port authority retains ownership in all shares of the venture. By applying market principles, the corporatized port authority is expected to function more efficiently. A corporatized port authority may also accommodate both national and local interests, as in the case in Poland. In the case of a publicly managed terminal, corporatization is usually the first step onto the road to privatization. Thus, a corporatized port authority, especially when based on a specific law, can be considered a permanent organizational structure while a corporatized terminal usually is a transitory organization

Corporatization, then, is the process in which a public sector undertaking, or part thereof, is transformed into a company under private corporate law. This is achieved by selling shares in a new company that conducts the port's business and holds its assets, although the shares are issued and may be owned entirely by the government (or port authority). The main objective is to decrease direct government control over the company and to make it more responsive to market forces. Similar to privatization, corporatization can include financial restructuring and be a catalyst for the introduction of commercial principles. Corporatization is, in effect, privatization without divestment.

For political or legal reasons (often both), comprehensive or partial privatization may be neither appropriate nor possible. In such cases corporatization may offer an effective alternative for achieving more efficiency and greater market orientation.

Corporatization usually features most of the following characteristics:

- A complete separation of the public management and regulatory functions from the commercial activities that are being corporatized.
- Clear and non-conflicting objectives for the new firm, set by the government.
- Greater management responsibility and autonomy for decisions on operations, investments, revenues and expenditures, and on commercial strategy.
- Where no market-based scrutiny is possible, performance measurement against a range of financial and non-financial criteria.
- Rewards and sanctions for managers based on performance.
- Government ensures that the corporatized firm does not have any comparative advantages or disadvantages relative to private port firms

operating under similar market risks and conditions (for example, with respect to tax and interest rates).

Corporatization can be implemented either through incorporation under a commercial code as a limited liability company or as a statutory authority under its own articles of incorporation. The statutory option is the most common approach for corporatizing port authorities. In view of the public interest involved, it is also the most appropriate one.

During the initial phase of the corporatization process, the following principal actions are required:

- Preparation and enactment of any needed legislation, such legislation often serves to eliminate the state monopoly within the affected sector.
- Development of the company charter (for example, the memorandum and articles of incorporation) for the corporatized port enterprise, and its subsequent incorporation.
- Development of a corporate plan including traffic forecasts, a business development plan, and pro forma income statement and balance sheet.
- Capitalization and vesting of part of the assets and liabilities of the former public company in the new corporation.
- Creation of a new labor statute, provision of financial and social measures to cope with excess personnel (such as pension fund guarantees, redundancy payments, or retraining), and transfer of personnel from the former public entity.
- Retraining of management and staff to increase commercial orientation and improve managerial procedures.

The key difference from the other reform options discussed is that the goal of corporatization is to constitute the corporatized firm as a single, self-contained entity. The corporatized company's management should be free from direct government interference or control (bureaucratic constraints) to allow them to operate the company on commercial terms. At the same time, management should also be held accountable for its actions. The new corporation can be organized with clearer lines of communication and responsibility. Distinct targets can be set and adhered to. Stricter internal financial controls can be introduced and, where necessary, information and accounting systems established. This all seeks to make the business more aware of market and client requirements.

One of the corporatized terminal's greatest strengths is its financial autonomy. This means that tariffs should no longer require approval from the government or ministry (unless it is a monopoly environment and the government wishes to exercise strict control) and that the company should be allowed to establish its own procurement, contracting, and hiring and firing practices. In addition, such companies do not rely on government support for investments and have the authority to negotiate loans directly with commercial banks. The government, however, typically will continue to exert some measure of political control. Usually this is achieved through the appointment of board members.

CORPORATIZATION OF A PORT AUTHORITY

Among the reasons for pursuing corporatization over other alternatives are:

- To allow time for the management to settle into its new role before contemplating full privatization (as is the case of the Rotterdam Municipal Port Management, until January 1, 2004, a commercialized port undertaking).
- To overcome the reluctance of private capital suppliers to invest in the company.
- To protect the public interest.
- Having completed the corporatization of port operational activities, subsequently one can consider the corporatization of the port authority as a regulatory body (for example, the case of the port enterprise of Antwerp). Negative aspects of corporatization include:
- In a majority of cases, the new corporate entity still has a monopoly over the port land.
- Unless competition is created, the corporation may not be as efficient as anticipated.
- Governments are still able to politicize the corporatized firm by retaining the right to appoint board members and executive directors.
- There will often be a need to introduce a port sector regulator to create a level playing field among competing service providers.

However, the most problematic issue affecting corporatized port authorities is the mix of public and private objectives. The rationale behind this type of reform is the expectation that corporatized ports operate as viable and effective businesses. However, while part of the ports enabling legislation may state that they should pursue commercial objectives and operate as effective businesses, the public shareholders (ministers, commissionaires, aldermen, or council members) have responsibilities other than strictly commercial ones, such as the delivery of public goods.

There are two types of corporatization models. The first model's goal is to transform former statutory authorities into government-owned enterprises. This means that a corporatized port authority would have a constitution consisting of a memorandum and articles of association that define the nature of the company and the manner in which the affairs of the company are to be conducted based on the companies act or corporations act in force. A regulatory body in existence should oversee performance of the newly formed port authority and ensure that conditions of the company's constitution and of the applicable companies act are met. This model has been applied to Rotterdam Municipal Port Management.

The second model involves the creation of a statutory government-owned enterprise (corporation) by specific legislation. This would mean that there is the potential for some degree of public (national, regional, or municipal) input and scrutiny. It also means the introduction of tailor made provisions, such as those relating to accountability and public control.

The distinction between the models focuses on the issue of whether the organization is subject to corporate law or to the conditions of the statue and specific legislation. The difference between a company incorporated under corporate law or by or pursuant to a statute is that the company's constitution spells out the nature of the company as well as regulations for the internal government of the company. This requires a rigid operating framework and a regulatory regime that ensures that the conditions of the company's constitution are neither breached nor abused to suit political or other gains.

Corporative port authorities established by law as government-owned enterprises, on the other hand, are quasi-private sector companies. They are expected to operate like their private sector counterparts, but are not subject to corporation's law, instead they are subject to the provisions of the statute under which they were enacted. Under this model, the public sector holds a pivotal role in the structure and operation of the organization.

Ultimately, the choice of one of the alternative models when corporatizing a public port authority is a political issue. In some countries, (larger) ports are considered part of the public domain, representing vital public interest. Other countries view ports mainly as commercial entities. The quality of governance also plays a role. Stable and democratic countries will be less inclined to corporatize their port authorities, unless for very specific reasons, which often have little bearing on efficiency. In Poland, the ports were corporatized to combine state and municipal ownership of port land. In Australia, the policy for port reform was an endeavor to improve efficiency in the port environment, notably by distancing government from day-to-day operations. Box 18 describes the process of corporatization for the Aqaba Container Terminal in Jordan.

PRIVATIZATION

Privatization can be either comprehensive or partial. The latter takes the form of a publicprivate partnership and is usually combined with the introduction of a landlord port authority. Comprehensive privatization remains an exception and is not a preferred option for major ports. The reasons that might prompt governments or a port authority to enter into the privatization process are discussed below. Removal of trade barriers. Outdated work practices, obsolete facilities, inadequate institutional structures, and excessive charges in ports cause inefficiencies that can create obstacles to foreign trade. Indirectly, the entire population of a country pays for port inefficiencies, which are reflected in the prices of both import and export commodities.

Harnessing the efficiency and expertise of the private sector. Increasing specialization in the shipping and port industry requires highly trained personnel, advanced systems and equipment, and capital-intensive cargo handling techniques to meet the fast changing demands of port users worldwide. Government-owned firms, with their cumbersome administrative procedures, poor cash flow generation, inflexible payment schemes, and lack of market orientation usually cannot cope with these requirements.

Elimination of political interference. Although there are countries with well-balanced political systems and minimal political interference in the functioning of the state- or municipal-owned port enterprises, the appointment of political nominees with inadequate experience to high level positions in government-owned ports is a well-known phenomenon. In contrast, privatization of port operations often results in the selection of professional port managers with an undiluted focus on the market and its changing needs.

Reduced demand on the public sector budget. Partial privatization does not necessarily mean a total withdrawal of the government from port investments. However, a large (often major) part of port investments can be undertaken by the private sector without compromising wider social and economic benefits.

Development of a modern port still requires a balanced public-private financial package with balanced risk sharing. Reduced expenditure on port labor. Government-owned enterprises traditionally have been a large source of direct employment; in the port sector, the greatest employment is in cargo handling services. A privatization scheme that maintains restrictive working practices cannot be effective. In the long run, creating an internationally competitive port system, with all its direct and indirect economic spin-off effects, is more valuable than the short-term objective of maximizing local dock labor.

Other objectives. Governments sometimes pursue privatization for other reasons, such as raising revenues for the state treasury, disposing of assets, and encouraging competition and broader citizen participation in share ownership.

In its many variations, privatization usually includes the following core features:

- Divestiture (selling off government-owned assets).
- Deregulation.
- Competitive tendering.
- Private ownership of operational assets with market-based contractual arrangements.

In theory, privatization provides the same flexibility to management as commercialization. Unlike under commercialization (where in the worst case scenario the government is likely to subsidize the company if it fails to perform adequately), a privatized terminal operation can be permitted to fail, provided other facilities can handle its traffic. Or, existing facilities may be taken over by a new operator who continues the operations. The management determines its own fate, free from significant government influence, as long as it complies with regulatory requirements.

REFORM TOOLS

Before deciding on a port reform process, governments should articulate clearly the ultimate goals of reform.

Broadly, there are two alternatives:

- The public authority in charge of the port sector (either a service port or a tool port) wants to restrict its public role by privatizing cargo handling operations and other non-landlord activities. In this case, existing operations have to be privatized or corporatized and service or tool ports reconstituted as a landlord port. Partial privatization is the goal.
- The public entity that has final responsibility for the port sector (most probably a national government) wants to privatize the entire sector, including responsibilities that generally are considered belonging to the public domain. Ownership of port land, planning, investment and management are all transferred to private sector entities, which have no formal commitments to any public institution. Comprehensive privatization is the goal (see Box 19 for an example of this type of privatization process).

This section focuses on the implementation of partial privatization, since that approach has been used successfully to balance public and private interests and still meet the objectives of port reform. Box 20 shows the spectrum of port reform tools that will be discussed in greater detail in this section.

CONTRACTING OUT AND USE OF MANAGEMENT CONTRACTS

One tool available to governments to improve port efficiency and performance is contracting out to the private sector certain functions previously executed by the public port management.

A public enterprise may decide to contract out certain of its operations through a tender-bid procedure instead of conducting them in house when the following circumstances apply:

- The functions can be performed at a price that is substantially lower than the cost of conducting them in the public sector.
- There is a large field for competitive bidding.
- Government policy is to transfer gradually certain non-core activities of the public sector to the private sector.

Contracting out, however, should be handled with caution as it involves several risks:

- If the number of potential bidders is limited, a meaningful comparison of the bids may not possible.
- Potential bidders may form a cartel or otherwise collude when bidding for a contract.
- Contracting out may create a monopoly for those activities, which would be contrary to the public interest, unless there is a proper regulatory oversight framework.

Also within the framework of commercialization, a separate contract for the management of the public port authority or public terminal operator may be awarded. Use of such a tool may be appropriate in cases where a port authority has experienced poor management for an extended period of time; the financial condition of the port authority needs to be substantially improved with a view to its corporatization or privatization at a later stage on terms favourable to the ministry of finance of the country concerned; or the port authority would generally benefit from the introduction of private management.

The usual practice is for the government to agree on a management contract with a private sector operator. The operator agrees to employ the existing port staff and to provide adequate and efficient service to all customers. This former requirement (retention of existing staff), however, often emerges as the main reason for the failure of management contracts (for example, the Port of Mombasa). The management company may be saddled with excess labor and labor costs that cannot be sustained in a competitive market.

A management contract is usually entered into for a specified period, generally between three and five years. Upon expiration of the contract period, it may either be renewed or awarded to another party. A management contract may also be used as a stepping stone towards the granting of a more extensive concession. It is important when entering into a management contract that the government or ministry has the right to impose financial penalties or terminate the contract in case the private operator does not meet specified minimum levels of efficiency, financial performance, or throughput.

CONCESSION ARRANGEMENTS

In concession agreements, governments are still widely involved in port management, mainly through public landlord port authorities. At the same time, the role of private enterprise in the sector will continue to grow. Service and tool ports will gradually disappear and be transformed into landlord ports; in some cases, fully privatized ports will emerge. For landlord ports, public bodies will retain the ultimate ownership of assets (especially land), but will transfer a major part of the financial and operational risks to the private sector. Governments will act mainly as regulators and land developers, while private firms will assume the responsibility for port operations. The main legal instrument used to achieve this realignment of public and private sector roles and responsibilities is a concession.

Concessions are widely used in the port sector today. A port concession is a contract in which a government transfers operating rights to private enterprise, which then engages in an activity contingent on government approval and subject to the terms of the contract. The contract may include the rehabilitation or construction of infrastructure by the concessionaire. These characteristics distinguish concessions from management contracts on one end of the reform spectrum and comprehensive port privatization on the other. Concessions, by permitting governments to retain ultimate ownership of the port land and responsibility for licensing port operations and construction activities, further permit governments to safeguard public interests. At the same time, they relieve governments of substantial operational risks and financial burdens.

There are two main forms of concession used in ports today: lease contracts, where an operator enters into a long-term lease on the port land and usually is responsible for superstructure and equipment, and concession contracts, where the operator covers investment costs and assumes all commercial risks. Such contracts are often combined with specific financing schemes such as BOTs.

Lease contracts and concession contracts share the same principal characteristics:

- The government or public port authority conveys specific rights to a private company.
- They have a defined term (10-50 years).
- They are geographically delimited.
- They directly or implicitly allocate financial and operational risks.

LEASEHOLD AGREEMENTS

Landlord ports derive a substantial part of their income from leases. Typically, only land or warehouse facilities are leased. Berths may be included or excluded from the lease rent. If excluded, the port authority collects and keeps all revenue derived from berthing fees. There are two basic forms of leases most commonly in use today: flat rate and shared revenue leases. Both types of leases can be used for multiuser as well as single-user (dedicated) terminals or berths.

Flat rate leases give the lessee the right to use a fixed asset for a specific period of time in exchange for periodic payments of a fixed amount. In the case of a land lease, this can be a fixed payment per year per square meter. Lease rates may vary depending on the degree of port site development (for example, unpaved versus paved land or land with or without structures). The main advantage of this form of lease is that the lease rent is known to both parties in advance. The flat rate lease also provides to the lessee the greatest incentive to fully use the available capacity of the terminal.

The main characteristics of the flat rate lease are:

- A specific sum of money is paid per square meter of port area for a specific period of time.
- In principle, the lease represents a fair return to the port authority on the value of the property.
- Lease payments may be adjusted for inflation over the life of the lease.

To set lease payments at the proper level, the port authority must be able to forecast accurately the level of business (and, hence, the wear and tear on port infrastructure and the traffic from which the lessee will benefit). It should also try to assess the true value of the land (for example, in its best alternative use) and attempt to recover this value through the anticipated level of business transacted by the lessee.

Because the lessee must make the same lease payment regardless of the revenue his business generates, he has a strong incentive to make full use of the leased land and structures. A flat rate lease is often the preferred form of lease for a port whose primary objective is to maximize throughput and benefits to the local economy.

In a shared revenue lease, the lessor also gives to the lessee the right to use a fixed asset for a fixed period in exchange for a variable amount of money. With a shared revenue lease there is a minimum payment regardless of the level of activity, but no maximum payment.

The main characteristics of the shared revenue lease are:

- A minimum level of compensation.
- No established maximum level.
- Maximum compensation depends on the facility's capacity.
- Minimum compensation may not fully cover the interest and amortization of the lessor (port authority) for the lease area.

A shared revenue lease represents true partnerships between the port authority and the lessees. Under this arrangement, the port must carefully determine the minimum lease payment, taking into consideration its financial obligations, its own forecasts of traffic volumes, and its statutory and business tolerances for risk. Once minimum throughput levels are attained, the lessee and the port share the benefits deriving from any additional activity. The shared revenue lease is the only approach in which the port authority can maximize revenues, employment levels, and throughput. Along with this potential for added rewards, however, come added risks.

Box 21 shows how the two different forms of lease would work for a notional terminal.

Potential lease partners for a port authority are:

- Terminal operators.
- Cargo handling companies.
- Dedicated terminal operators and shipping lines.
- Forwarding agents.
- Inland transport operators.

Today it is increasingly common for shipping lines to lease terminals from port authorities. For these leases to succeed for all parties, however, two key conditions should exist: the shipping line lessee should generate a large volume of cargo at the port (that is, it should be a major customer), and the port should possess additional facilities of the same type leased to the shipping line to prevent creating a monopoly (a public access facility should be available).

If the port does not have other similar facilities (and other customers), the creation of a monopoly may conflict with the interests of both the port and the national economy.

In this respect, the following points should be kept in mind:

- Shipping lines may, at any point in time, decrease, reroute, or altogether halt their services as a result of changes in financial conditions or shifts in patterns of trade. A well-known example of this is the cancellation of the round-the-world service of United States Lines in the 1980s.
- Shipping lines often merge or enter into cooperation agreements (alliances) with other shipping lines. Such practices may result in changing sailing schedules or the establishment of special ties with other ports.
- Shipping lines may reorganize their sailing schedules for reasons of internal policy.

Signing a lease contract with an operating company may be less risky than with a shipping line because the operating company usually does not rely on a contract with one single user, but will spread the risks and safeguard its business interests by having contracts with several clients, and in the case of a contract with a locally incorporated port operator, should a legal (contract) issue arise, it is generally easier to enforce liens and other measures needed to compel lease compliance than in the case of a company whose home base is in another country.

Which form of lease is to be preferred? In general, one may conclude that if the ports principal objectives are to maximize throughput and provide maximum benefits to the local economy through increased employment, a flat rate lease may be preferable. This is often the case when a port is newly established and wants to develop its business. Or if the ports principal objective is to maximize revenues, with an initial need to subsidize the terminal lessee, the shared revenue lease may be the optimal choice.

Concession Agreements

A landlord port for the most part does not involve itself directly in port operations. Instead, private port operators and service providers conduct their business independently and compete in the market. The port authority acts as a neutral landlord promoting the port as a whole. Together, they represent the interests of the entire port, with the port authority in the lead.

Relations between the port authority and the private sector cover two areas: commercial relations based mainly on concession and lease agreements, and relations based on the public oversight functions of the port authority, such as enforcement of port bylaws, dangerous goods regulations, and vessel management.

Relations between landlord port authorities and private port operators have become increasingly complex, and the alignment of responsibilities have further shifted. One of the valued features of a landlord port is its clear division of responsibilities. Each party is distinctly aware of its rights, liabilities, and financial responsibilities.

Moreover, many governments today are seeking to diminish their financial involvement in ports and to use private sources to finance new port development, including construction of basic infrastructure such as quay walls. This implies not only an increased role for the private sector in port development, but also increased financial exposure. In such situations, a simple and straightforward lease contract often is not sufficient to cover all responsibilities and liabilities. As a result, a more complex contractual relationship, a concession agreement, has been developed.

The primary objective of concession agreements is to transfer investment costs from the government to the private sector. Concessionaires are obliged to construct and rehabilitate infrastructure and operate a facility or service for a fixed number of years. Concessions may be positive, when a concessionaire pays the government for concession rights, or negative, when the government pays a concessionaire for the services it provides under the agreement.M

The benefits of concessions in the port sector include:

- Better and more efficient port management (especially port operations) performed by private operators.
- Avoidance of the drawbacks associated with monopolies through the inclusion of detailed concession conditions.
- The application of private capital to socially and economically desirable projects, freeing up government funds for other priority projects.
- Under certain circumstances, the creation of new revenue streams for governments.
- The transfer of risks for construction, finance, and operation of the facility to the private sector.
- The attraction and use of foreign investment and technology.

Disadvantages associated with concession contracts include:

- The need for continuing close government regulation and oversight.
- The system requires a legal framework that permits transfer of land rights to a private party.
- Winning bids are sometimes based on unrealistic financial projections, placing the sustainability of the concession agreement in jeopardy.
- The danger that a concessionaire will not properly maintain the facilities under concession, returning them to the government in bad condition, or the danger that the concessionaire and the port authority disagree on the operational need for and financial feasibility of critical investments.

Concession agreements are often developed as a part of a BOT scheme and represent specific agreements between a government or port authority and the special purpose company (SPC) established by the concessionaire to carry out construction and operation of a port development project. Under concessions, the ultimate ownership of the affected assets is retained by the national or local government, or by the port authority. At the same time, part of the commercial risks of providing and operating the assets is transferred to a private concessionaire.

In agreements involving an SPC, a port authority should ensure that:

- The SPC provides adequate service throughout the term of the concession.
- The SPC observes relevant safety and environmental protection standards.
- The charges levied on port users are reasonable and do not endanger the competitive position of the port.
- The SPC performs proper maintenance and repair of all assets to ensure that on their return at the end of the concession, the port authority receives an operational project and facilities in good working order.

The port authority may (depending on legal strictures) hold a financial interest in the SPC created by the concessionaire, or it may not. If the port authority chooses not to participate financially in the SPC responsible for developing the port assets under a concession contract, then its role as an independent and impartial public entity does not significantly change. The only real change is in the shift in responsibility for investments from the port authority to the concessionaire.

If a port authority not only enters into a concession agreement with the SPC, but also participates in the company as a shareholder, then the port authority's role changes more dramatically. By investing risk capital, the port authority becomes more directly involved in port operations. Sometimes this situation is prohibited by law (Poland). If the venture has a monopoly in the port (such as having the only container terminal), the situation might be acceptable, although a conflict of interest may arise between the roles of port authority as an investor and as the regulator of the monopoly. If the venture competes with other terminals in the port, however, participation of the port authority in the SPC will give rise to a serious conflict of interest and will undermine its independent, neutral position.

Depending on the specific situation, a concession agreement may consist of a combination of contracts including:

- A leasehold agreement on nondeveloped land, the formal document under which the port authority grants the SPC possession of the concession area.
- A terminal access agreement, which regulates the SPC's access to the concession area, and also the access by the port authority to the area.
- A port services agreement, which regulates the provision by the port authority to the SPC of various port services such as pilotage, towage, and dredging.
- A sponsor's direct agreement, which is an agreement between the government or port authority and the SPC dealing with the issue of competition.
- A design contract between the SPC and a technical consultant for the design of new facilities (the port authority usually has no direct control over who does the design work or the terms of appointment, but often retains the right to review any design).
- A building contract between the SPC and a construction company for construction or development work (with the port authority typically exercising some form of quality control).
- Financing documents drawn up between the SPC and its lenders to provide finance for port development; a port authority may provide partial financing.
- A management contract between the SPC and its chosen manager (operating company) for provision of management services in operating the port.

Generally, a typical concession agreement will clearly set out the terms relating to:

- The land, facilities, and cargo handling equipment included in the concession.
- The functional requirements of the port or terminal, the proposed design solution for any construction, the construction programme, and time schedule, including milestones.
- Rights and responsibilities of the concessionaire and port authority (concession sponsor) with respect to the completion of the construction programme.

- Human resources development and the employment of former port authority employees, if applicable.
- Activities permitted to be carried out in the concession area.
- Equal access to common areas in the port.
- Payment of fees, royalties, revenues, and canon (lease rental) to the port authority.
- Maintenance requirements for infrastructure, superstructure, and sometimes equipment.
- Termination of the concession.
- Return of land, facilities, and equipment after the concession period has expired.
- Other issues as may be required.

It is common practice that during construction, the concessionaire and the port authority use an independent test certifier to certify that all work has been carried out in conformity with the requirements of the concession agreement. Upon the return of facilities, the SPC should be required to carry out any work needed to bring them up to an agreed-on standard. Accordingly, provisions must be included to inspect facilities and identify any deficiencies.

A concession agreement for a greenfield project is less complicated than the takeover of an existing terminal or port. In such a case, no personnel or existing facilities are acquired by the SPC. However, a terminal access agreement still must be drawn up between the government or port authority and the SPC to cover such things as the building of access roads and rail, the provision of water and electricity, and other facilities.

Master Concession. In some instances, port reform is implemented through a master concession contract, which enables a private operator to carry out many of the port functions. This type of contract has rarely been used, but it is an option. Usually, the principal choice is between granting a full master concession, in whatever form, and implementing a landlord port structure comprising the public port authority and private terminal operators. The choice between the two options considerably influences further port privatization process. When choosing a master concession, the government leaves the unbundling of port activities for a large part in the hands of the concessionaire. It might also be expected that retrenchment costs resulting from granting a concession would primarily be borne by the government.

The government should allow the concessionaire enough freedom to structure its business according to its own requirements, otherwise the exercise does not make much sense. Lack of freedom will lower the concession's attractiveness. To make a master concession attractive for a private investor, the concessionaire should be allowed to unbundle the port business in the way it thinks fit. On the other hand, introducing a landlord port system will require a much more active role for the government in structuring the various concessions of terminal and marine activities, as well as reorganizing the port authority.

BOT Arrangements. A landlord port authority is typically responsible for constructing fairways, quay walls, and terminal areas. Such construction is usually based on a port master plan and carried out in close consultation with the future operator. Sometimes construction of such facilities has already started before agreements have been concluded with the prospective operators. This may be the case when the market demand is strong and the port authority is confident of finding clients and is prepared to take the risk that port capacity will go unused. As a rule, port authorities should permit private operators to finance most of the additional capacity (including the quay wall expansion). The port authority can then concentrate on access infrastructure and protective works relating to port extension and on renovation projects. Port authorities may sometimes have difficulties amassing the investment funds from dues or retained profits. In such cases, they have sought to acquire funds either from an IFI (such as the World Bank) or from private lending institutions. For specific port facilities, such as container or bulk terminals, private funding can be arranged through a concession agreement as described above. BOT schemes are a specialized form of concession designed to increase private financial participation in the creation of port infrastructure and superstructure without changing the landlord structure of the concerned port (see Box 22).

When designing BOT schemes, it is important to consider carefully which parts of the port can be concessioned and which parts should remain with the port authority. Generally, BOT schemes can be applied to all assets that can be exploited as a separate business.

Key among these are:

- *Fairways and channels:* This part of the port infrastructure can be concessioned under a BOT scheme to require the concessionaire to dredge and maintain the fairway (and, optionally, to operate aids to navigation) for a specified period during which it derives an income from vessels using the fairways under an agreed fare system (for example San Martin-Rosario Fairway, Argentina, described in Box 23).
- *Terminals:* BOT schemes are usually applied to specific terminals. There are many examples of such terminals, such as the former P and O terminal at Nhava Sheva, India; the South Asia Gateway Terminal at Colombo; the Aden Container Terminal; and the Port of Buenos Aires, Argentina.
- *Entire port complexes:* A BOT structured as a master concession contract could cover an entire port complex comprising various terminals. Here, the SPC (or port operator) assumes de facto the role of a landlord port authority for the assets it has agreed to

construct. The master concessionaire then offers subleases of various terminals to third parties. Such a scheme can approach comprehensive privatization. The only real distinctions are that under a BOT and master concession, the transfer of assets is temporary and the concessionaire has no regulatory responsibility for marine safety, environment, or vessel traffic management. There are no examples of effective implementation of this type of BOT master concession scheme, but new legislation in Madagascar provides for une concession globale, which is the equivalent to a master concession for small ports of local interest.

Other port assets cannot be easily concessioned as individual items. The most important of these are assets such as breakwaters, piers, connecting channels, intraport roads, and other common areas. These assets, however, can be part of a master concession agreement or a comprehensive privatization scheme.

A carefully crafted concession is central to the implementation of a BOT scheme. The concession contract gives the concessionaire the right to run the facility (with limited and clearly defined government oversight) and earn a commercial return on investment. The concession or BOT agreement, with the required business plan, will set out estimates of the likely revenues, costs, debt repayment, and profit for the SPC. This information is necessary to assess the project's financial viability and its debt repayment capacity. Many planned BOT projects fail because their terms are negotiated without taking into account whether or not the project is bankable. Governments often try to negotiate a BOT arrangement at an early stage in the project preparation cycle, before the full scope of the project is known and before a regulatory oversight regime has been decided. While this might generate significant revenues for the government in the short run, it may saddle the concessionaire with an impossibleto- complete project.

There are many variants of BOT-like schemes, including:

- *Build-own-operate (BOO):* Full privatization of the terminal because the port land and the facilities built on it are not returned to the government or port authority.
- *Equip-operate-transfer (EOT):* Port infrastructure already exists, but superstructure is supplied by the SPC.
- *Build-transfer-operate (BTO):* New port facilities are directly transferred to the competent authority (government or port authority) immediately after construction. Under BTO schemes, the ownership of the assets being financed has been an issue for lenders who require asset-based collateral to secure bank loans. With BTO schemes, the only collateral is the concession contract itself, which may be insufficient. BTO schemes are necessary in countries where legal

strictures do not permit private ownership of main port infrastructure (for example Croatia, Italy, Costa Rica, and the Republic of Korea).

- *Build-own-operate-transfer (BOOT):* Ownership of land and facilities conveys to the concessionaire, but is transferred back at an agreed-on price at the end of the concession period.

A special case is the wraparound BOT (WBOT); this scheme is used in the case of expansion of a government-owned port facility by the private sector, which would hold title to the expansion only.

Under such a scheme, the SPC would:

- Operate the entire port facility under a project development agreement.
- Manage the government-owned section under a management contract.
- Expand the facility under a BOT contract.

In many cases, the government effectively becomes a partner in a BOT arrangement by investing in certain portions of the infrastructure. Private parties appear to be reluctant to invest in basic port infrastructure, not only because it makes it more difficult to price use of infrastructure in a manner that permits the concessionaire to realize a reasonable return on the investment, but also because these assets are largely immobile and have no comparable alternative use. Political instability, change of control, antiprivatization backlashes (nationalization), unexpected new tax regulations, and other governmental actions could make comprehensive BOT schemes much less attractive.

COMPREHENSIVE PRIVATIZATION

Comprehensive port privatization has, until now, been developed only in the U.K. and in New Zealand. Outright sale of port land combined with a transfer of traditional public port tasks, such as safety and environmental oversight (for example, harbormaster's tasks), remains an exception. Other countries have introduced significant privatization schemes, but mostly with respect to port and terminal operations. Comprehensive port privatization often requires the enactment of new laws, both to regulate the transfer of ownership and functions from the public to the private sector and to define the borderline between redrawn public and private responsibilities and tasks.

Such legislation should establish:

- Authority for the port authority to establish a new successor company or companies to take over all or part of the authority's business.
- The right of the successor company to issue shares, either to the authority or to a third party.
- The time and manner for selling or otherwise distributing the shares to third parties, as well as for a payment to the successor company from the proceeds of the sale.

- The basic authority and mechanisms needed for the government to shape and direct the privatization.
- A levy on the proceeds of the disposal of shares of the successor company (in the U.K. this levy was set at 50 percent of the net proceeds of the sale).
- A levy on profits accruing to the successor company as a result of the disposal of port land transferred under the privatization scheme (in the U.K. this levy was set at 25 percent of the profit during the first five years, 20 percent during the next two years, and 10 percent during the last three years of the levy period).
- Provisions for the transfer of port authority personnel to the successor company (for example, the number and categories of personnel, salaries, benefits, and pension rights) or their dismissal (for example, separation package, retraining allowance, rehiring preferences).
- Terms for the transfer of public tasks, such as aids to navigation, pilotage, handling of dangerous goods, and protection of the environment to the successor company or other entity.
- The tax regime applicable to the successor companies.
- Authority for the government to dissolve the port authority once it is satisfied that the objectives of the enabling legislation have been met and to transfer all remaining property, rights, and liabilities to the successor company.

Privatization legislation may include additional elements, depending on the local situation, the structure of the former port authority and the specific legal, institutional, and socioeconomic situation in the country concerned.

In the U.K., the benefits of comprehensive port privatization most often cited are:

- The generation of revenue for the treasury.
- The ability of privatized companies to diversify their businesses.
- Greater access to capital markets.
- The removal of restrictions on investment and borrowing.
- The introduction of new industrial relations practices.
- A more commercial and entrepreneurial approach to management of the business.
- Greater competition.

These features, it was argued, would result in improvements to the port system's financial and operational performance.

Note, however, that not all of the above-mentioned benefits are due exclusively to comprehensive privatization; other port reforms may generate similar benefits.

A vast majority of maritime nations considers comprehensive privatization to be incompatible with national and regional interests.

Specific reasons why governments and port authorities have refrained from pursuing full privatization are diverse, but often include one or more of the following:

- A public monopoly can easily become a permanent private monopoly.
- The macroeconomic benefits of large port complexes to the regional and national economy are perceived to be threatened by comprehensive privatization.
- The danger of discriminatory treatment of customers.
- The risk that, in practice, privatization may undermine competition.
- Fear of overinvestment in and duplication of dedicated terminals for major clients, which could unbalance demand for additional public transport infrastructure.
- Neglect of the port's public service function.
- Reluctance of labor unions to abandon government protection and their fear of losing jobs.
- Reluctance of public authorities to lose political control, including patronage.
- Reluctance of public authorities to lose income generated by the port business.

Background on the U.K.'s port privatization is provided in Box 24a and 24b. After more than 10 years of experience, some conclusions can be drawn concerning the U.K's implementation of comprehensive privatization. Generally, the U.K. model of port privatization is highly determined by local factors and ideological considerations that are unique to the British experience.

However, it appears that:

- The valuation of port assets sold to private parties was judgemental because there was no established market during the time of privatization.
- Subsequent trading of port shares suggests that the original prices were only 25 percent of their true market value.
- Ports were sold at significantly discounted prices. Discounted sales (in addition to the ruling that 50 percent of the sale proceeds from disposal of Trust Ports should be returned to the buyer) significantly reduced the original debt of the new port company. Certain privatized Trust Ports, therefore, realized very high profits (as high as 20-30 percent of turnover) at the expense of port users and taxpayers. Although difficult to prove, privatization via a concession, rather than outright sale, would probably have raised considerably larger revenues for the public treasury.
- Transfer of port regulatory functions to the private sector has raised serious issues. The new privatized ports are essentially self-regulating and have little incentive to safeguard and enhance interport competition. The driving force behind the new port owners is

corporate interest rather than public interest. The question, then, is who protects the public interest?

- In terms of investments and profits, privatized U.K. ports have done better than the still-existing public ports. Privatization led to an injection of cash, but only for purchasing existing assets. Former Trust Ports claimed that investments were hampered by financial institutions looking only for short-term returns.
- The valuation of port assets sold to private parties was judgemental because there was no established market during the time of privatization.
- Subsequent trading of port shares suggests that the original prices were only 25 percent of their true market value.
- Ports were sold at significantly discounted prices. Discounted sales (in addition to the ruling that 50 percent of the sale proceeds from disposal of Trust Ports should be returned to the buyer) significantly reduced the original debt of the new port company. Certain privatized Trust Ports, therefore, realized very high profits (as high as 20-30 percent of turnover) at the expense of port users and taxpayers. Although difficult to prove, privatization via a concession, rather than outright sale, would probably have raised considerably larger revenues for the public treasury.
- Transfer of port regulatory functions to the private sector has raised serious issues. The new privatized ports are essentially self-regulating and have little incentive to safeguard and enhance interport competition. The driving force behind the new port owners is corporate interest rather than public interest. The question, then, is who protects the public interest?
- In terms of investments and profits, privatized U.K. ports have done better than the still-existing public ports. Privatization led to an injection of cash, but only for purchasing existing assets. Former Trust Ports claimed that investments were hampered by financial institutions looking only for short-term returns.
- The abolition of the National Dock Labor Scheme had a more profound effect on labor stability than the selling of port land.
- Where terminals were already privately operated (landlord ports), selling the underlying port land made little difference. For example, port land at Dover (a former Trust Port) or Portsmouth (a municipal port) did not affect port output because port operations in both ports were already in private hands.
- Some nationalized and Trust Ports were sold under a M(E)BO scheme to former public officials. These managers reaped windfall profits by selling their shares at a later date.

- There are limited possibilities for port cities to redevelop obsolete port land. On the other hand, land speculation by privatized ports has become a reality because older port facilities are often situated near the valuable real estate of city centers.

The U.K. experience, therefore, has yielded very mixed results and provides few arguments supporting comprehensive privatization (the sale of port land and transfer of all public functions to the private sector) when other, less radical reforms can achieve the same objectives.

PORTS AS TRANSPORT CHAIN FACILITATORS

Increasingly, major terminal operators are trying to secure their strategic position by offering complementary terminal facilities located either in the foreland or hinterland. This practice is most apparent in connection with containerized cargoes. In the event that an operator engages in operating other facilities such as inland terminals, rail facilities, or even entire port complexes abroad, its objectives and motivations are broader than those of a localized operator.

The phenomenon of supply chain management can for instance be well observed in the Port of Rotterdam, where very large crude carriers (VLCCs) discharge crude oil from various oil producing countries. Rotterdam has a virtual monopoly in this traffic in Northwestern Europe as a result of its very deep access channel to the North Sea (78 feet). Pipeline systems have been constructed to connect the port with various refineries in the hinterland, such as in Belgium and Germany. Thus, the inland transport chain is effectively controlled by one port, creating a stable environment for the transport of crude oil as well as an attractive location for balancing refineries. The Rotterdam Municipal Port Management was instrumental in developing the pipeline systems.

Some port authorities also seek to attract customers to their port facilities by facilitating or cofinancing terminal facilities outside their port area. This more expansive view of a port authority's role has the potential to influence traditional port management structures, particularly in ports structured on the landlord model. A port authority's involvement in terminal operations beyond its homeport may not be focused solely on improving logistics chains. The main objective might be to maximize the port authority's revenue by making more widespread use of its operational expertise and management, especially in the case where the port authority acts as terminal operator as well.

Port authorities seeking to become transport chain facilitators should be aware of possible conflicts of interest and the potential loss of their neutral position. Managing a port area, including attendant public functions, is different from optimizing a logistics chain, which can be considered a supporting function for the ports industry, and for that reason essential from a competitive point of view. The PSA Corporation is a prime example of globalization of terminal operations. Since its establishment, it has become a leading player in the global terminal operating business and today owns, manages, and operates a chain of container terminals and logistics hubs throughout the world. Before taking on this expanded role, PSA had to change thoroughly its legal structure. Box 25 describes this transformation.

MARINE SERVICES AND PORT REFORM

This section discusses a variety of marine services and how they are affected by port reform. Special emphasis is placed on how these services might be outsourced, concessioned, or privatized.

Marine services are port-related activities conducted to ensure the safe and expeditious flow of vessel traffic in port approaches and harbors and a safe stay at berth when moored or at anchor. Safe means that port conditions ensure that vessels using the port, the port environment, and the marine environment are protected from danger. Expeditious means that vessels are not unduly delayed and that the vessels port transit times, as a part of the total turnaround time in the port, are kept to a minimum.

Although ports may define marine services differently, and may have different methods of providing them, in this section the term is used to refer generally to services having a nautical bearing, be it maritime safety, vessel traffic efficiency, or marine environment protection. Other services (for example, fire fighting, immigration and customs services, security, and port state control) may also affect port efficiency and safety. While important to the overall operation of a port, these other services are not dealt with in this section.

The specific marine services rendered by a port authority depend largely on the scope of the port's marine responsibilities and jurisdiction. The scope of the ports marine jurisdictions does not follow a general rule, and there exists no international legislation or standard practice that defines the responsibilities of port authorities. Usually, marine services rendered by a port authority are geographically delimited by the area directly under control of the authority, which may encompass only the waterfront of riparian berths (the port's domain). However, there are countries where the port authority is also responsible for managing lighthouse services outside its immediate area of control. This extended area may cover harbor waters and approaches as far as the open sea.

HARBORMASTER'S FUNCTION

Generally, the harbormaster (or port captain) manages port activities relating to maritime safety and the protection of the marine environ-ment. The legal basis of the harbormaster's function is usually embedded in a port bylaw or, in the case of a state-owned port, in a specif-ic law or ministerial decree. The harbormaster often has specific legal powers to act in emer-gency situations. Typically, the harbormaster is part of the port authority organization and heads the marine department. In some coun-tries, the harbormaster may work for an inde-pendent public entity such as the coast guard. The harbormaster is responsible for ensuring the efficient flow of traffic through port and coastal waters (including allocation of vessels to public berths) and-on behalf of the govern-ment or port authority-for coordinating all marine services. The harbormaster operates out of a port coordination center (or Captain's Room), which is often part of an elaborate ves-sel traffic management system.

Frequently, harbormasters have police powers and act as head of the port police. The main functions of such police are enforcement of the port bylaws, especially with respect to traffic regulations, protection of the environment, and accident prevention. When part of a port authority, the harbormaster also usually serves as head of the pilotage service. In the event that the pilotage service is not part of the port authority, the harbormaster is responsible for coordination between this service and port users. Finally, the harbormaster is sometimes responsible for regulatory oversight of the car-riage and storage of dangerous goods in the port area as well as for ensuring the proper use of port reception facilities.

In view of the public character of the harbor-master's responsibilities, this function is rarely privatized. To do so would raise a conflict of interest between the public interest (safety, envi-ronment, and equal treatment under the law) and private interests from the port industry. For example, since port time of ships is an impor-tant cost and operational factor, the harbormas-ter will always be under pressure to grant pref-erential treatment to shipping lines. Impartial and consistent application of operational safety measures for ships carrying dangerous or envi-ronmentally sensitive goods such as gas carriers, chemical parcel tankers, and VLCCs is essential to the safe functioning of any port. The harbor-master, therefore, should not function within a purely commercial environment, but must have freedom of action to carry out public tasks in an unimpeded and unbiased manner.

Although the harbormasters might be part of a port authority's management team, they should be free to operate in their jurisdiction as inde-pendently as possible from the commercial management of the port. In carrying out emer-gency measures in the event of accidents and industrial disasters, the harbormaster should have full freedom of action and possess the ulti-mate authority and responsibility for directing all necessary activities. In a fully privatized port, the harbormaster should not be part of the port management, but should be employed by a national or regional maritime administration.

PILOTAGE

In a port reform process, pilots often are the first ones to demand privatization. Pilots usual-ly constitute a closed group of professionals (often master mariners), who are keenly aware of their unique position in the port environ-ment. Successful vessel management relies heav-ily on the efficient functioning of the pilot organization, a fact that pilots may use to maxi-mum advantage during port reform.

In many countries, pilots (or pilot organiza-tions) have been more or less successfully priva-tized. This type of privatization, however, car-ries the risk of creating a private sector monop-oly in pilotage services, especially when pilots are privatized on a national or regional scale. Pilotage is an essential part of traffic manage-ment, and safe passage of vessels through a port area requires expert teamwork of a vessel traffic management organization (Captain's Room), tugs, mooring gangs, and pilots. A private sec-tor pilot monopoly that has the ability to bring port operations to a complete and rapid stop represents a significant risk for ports, carriers, and shippers alike. As a consequence, retaining pilots as part of a port authority's marine department may be desirable even when other aspects of port management and operations are privatized (see Box 26).

There are two ways of privatizing of the pilotage function. Pilots can be self-employed and work under the oversight of a maritime authority that serves as the regulator and licen-sor of the individual pilots, or pilots can organ-ize themselves into a private company.

The pilotage company should have its own infrastructure and facilities, such as pilot boats, communication equipment, and pilot stations. Sometimes a pilot organization (especially in smaller ports) might also operate a vessel traffic management system (radar).

The port authority or maritime administration should regulate the privatized pilot organization regarding:

- Training requirements and pilot qualifications.
- Standards for obtaining a certificate or license, and its revocation.
- Roles and responsibilities of the organiza-tion for operation of a vessel traffic man-agement system.
- Communication equipment and channels.
- Investigation of incidents and follow-up actions.
- Pilotage tariffs and financial record keeping.
- Medical fitness and continued proficiency.
- Reporting requirements to the relevant port authority.

TUGBOAT OPERATIONS

Tugboat operations are typically carried out by private firms. If the volume of vessel traffic is not sufficient to support a tugboat service on a com-mercial basis, a port authority may be obliged to provide such service itself. Sometimes neighboring ports can share tugboat services to reach volumes sufficient to sustain a commercial operator.

In many instances, traffic density allows for only one private tugboat company to operate in the port area.

In such cases, the port authority should regulate the service regarding:

- Minimum crew size.
- Minimum bollard pull.
- Communication equipment and channels.
- Roles and responsibilities relating to the vessel traffic management system.
- Tariffs.

The optimum situation would be a number of tugboat firms competing vigorously in the port. In that event, the port authority should not have to regulate tariffs. Regulation of other aspects of tug operations such as manning can be at the discretion of the port authority and will depend on the local situation.

MOORING SERVICES

Mooring services in smaller ports can be pro-vided by the local stevedore. In larger ports, a mooring service is usually performed by a spe-cialized private firm. Especially in a complicated nautical situation (for example, single point mooring buoys, specialized piers for chemicals or gases, or ports with large tidal differences), mooring activities require expert skills and equipment. A port authority may choose to reg-ulate this activity when only one specialized firm exists.

Regulations should include:

- Minimum manning requirements.
- Communication equipment and channels.
- Number of mooring boats and their characteristics.
- Tariffs.

VESSEL TRAFFIC SERVICES AND AIDS TO NAVIGATION

Vessel traffic services (VTS) are usually part of a port or a maritime authority. Such services are provided in port areas and in densely used maritime straits (such as the Dover Channel) or along a national coastline (for example, the coast of the Netherlands). In principle, it is possible to privatize VTS under a concession agreement.

VTS that should be regulated by the competent authority should include:

- System functions, such as vessel manage-ment and control, emergency functions, and information and communication functions.
- Types and specifications of radars and tracking software.
- Manning levels and qualifications.
- Reporting duties.
- Tariffs.

Responsibility for aids to navigation usually rests with a national maritime authority in port approaches and in coastal areas, and with a port authority in port areas. Often, provision and maintenance of buoys and beacons are contracted out. Because aids to navigation are generally part of an integrated maritime infrastructure, the costs of providing these services are included in the general port dues. Therefore, it is difficult to privatize them.

OTHER MARINE SERVICES

The control of dangerous goods for maritime cargoes is usually performed by a specialized branch of the port authority. The same goes for the handling of dangerous goods in port termi-nals. Oversight and regulation of land transport of dangerous goods is normally a responsibility of the central government. The highly sensitive and technical nature of this work makes it inad-visable for privatization.

Waste management services in ports often are privatized under strict control of a port authority or another competent body. Privatization carries risks, however, especially with respect to the disposal of dangerous chemicals. Proper waste management can be expensive for shipping lines. With high costs, ship captains might be tempted to dump waste into the sea or into port waters. Control of such dumping practices is extremely difficult, especially for chemical cargoes. To spread waste management costs, ports can include all or part of the waste management costs in the general port dues. Transport of waste from the ship to a reception facility also poses a challenge, especially in larger port areas. Port authorities should directly provide or organize the provision of transport barges or trucks for this purpose.

The entire waste management system, including personnel and facilities, should be closely controlled by the competent authority. When private firms are engaged in waste handling, the authority should employ experts from its organization to ensure compliance with all relevant laws, rules, and regulations.

Generally, emergency response services are carried out by a variety of public organizations such as the port authority (harbormaster), fire brigade, health services, and police. Some ports have sophisticated tools available to aid in crisis management, such as prediction models for gas clouds. Such tools are often integrated in a traf-fic center of the local vessel traffic management system (VTMS). Private firms (for example, tug-boat companies) may play a subsidiary role in crisis management in the event that they are equipped with fire-fighting equipment. Larger ports use patrol vessels and vehicles for a variety of public control functions. In some ports, such patrol vessels also have fire-fighting equip-ment on board. When a port does not have patrol vessels available, a contract with a tugboat company should be arranged to guarantee avail-ability of floating fire-fighting capability. Port patrol services are part of the harbormaster's resources and, therefore, should not be privatized.

Control of dredging operations by a port authority is of utmost importance. Often, the port authority or the competent maritime administration does not have enough expertise to exercise sufficient control over both mainte-nance and capital dredging. Port authorities with large water areas under their control should employ sufficient competent personnel to prepare dredging contracts and oversee dredging operations. Sounding is an activity that should preferably be carried out (or contracted out) by the port authority itself. Dredging is usually carried out by private firms. It might be cost effective for some ports to use their own dredges, especially when continuous and important maintenance dredging is required.

PORTS AND ECONOMIC CHANGE

A port generally offers a value proposition to its regional since it offers economic and social benefits, but is also prone to environmental constraints. Significant increases in port throughput, particularly in the containerized sector, have put pressures for the development of new port infrastructures on existing facilities, and also for entirely new developments when additional capacity cannot be developed on existing sites. Ports are capital intensive infrastructures that are associated with a wide array of economic impacts. Port development and world trade are closely interrelated. There are numerous expectations by the public sector, which is often providing substantial capital investments (through the port authority or general funds), to see concrete and measurable economic impacts and benefits resulting from these investments. However, the existing literature is relatively scarce about the formal impacts of ports on regional development. Evidence is usually related to a single port over a narrow range of impacts, which makes general assessments difficult to make.

Economic impacts concern the wide range of changes brought buy infrastructure investment projects while economic benefits tend to be directly measurable impacts in terms of a monetary value. However, many of these impacts can only be observed after the investments have been made and the benefits measured. An ex-ante (forecasting) exercise is hazardous and commonly lead to inaccurate assessments. Port forecast models are rarely accurate. The bottom line is that the estimation of economic impacts of port investments is an inexact field, which focus on the effectiveness of transport infrastructure as a catalyst of indirect and induced benefits. Further, these investments are contingent to the scale and scope of changes in which they are taking place.

Among the most relevant changes that have impacted ports and maritime transport:

- Economic changes. Seaborne trade has increased substantially, in part because of the massive redistribution of manufacturing to low cost locations (outsourcing) and in part because of ongoing economic growth. This underlines the growing importance of logistics to organize the resulting complex distribution system.

- Technical changes. The growth in ship size to better achieve economies of scale has been a prevalent technical change, particularly since the 1990s when post-Panamax containerships were first introduced. There is also a growing level of ship specialization (containerships, bulk carriers, car carriers, and even cruise ships) that required dedicated port terminal facilities. All of the above has been placing pressures on ports to upgrade and improve their facilities.
- Organizational changes. The maritime and port industry are increasingly controlled by large shipping companies and terminal operators that have engaged in strategic alliances as well as mergers and acquisitions. Their goal is to provide a level of vertical and horizontal integration, which is improving the performance of the port transport chain. This has led in a number of ports to the setting of inland terminals.

The outcome of these changes have involved port developments that are more capital intensive, while relying on less labor and consuming more land. The imperatives of maritime shipping companies have been felt on ports as they increasingly tend to compete to attract traffic, particularly since hinterlands tend to be more contested. The industry is expecting lower tariffs and lower port times in light of a highly competitive environment and low profit margins. Ports acting in a monopolistic fashion are finding themselves with less leverage, with negative impacts on their activity and regional economies.

The spatial framework of the port is also changing. Many port areas have seen the relocation of port industries to new sites, either within the region or to another country altogether. These changes have been associated with a dislocation of the relationships of many ports with their localities and regions; this has been labeled as port regionalization. While the port remains a strategically important infrastructure, its economic benefits are less directly apparent within the community with weaker but more complex relations at the regional versus global levels.

The impacts of port infrastructure investments are expect of a positive influence of port throughput on local economic development. However, evidence across the world underlines that this influence is weak, with elasticity levels between throughput and employment that are typically less than 0.05 jobs per 100 tons. This implies growth in traffic volumes are not associated with significant direct gains in employment. This elasticity is among the weakest in the transport sector, particularly in regard to airports, which are the infrastructure with the highest elasticity. Still, the employment impacts of ports are positive and are usually higher for the service sector than for the industrial sector. Empirical evidence underlines that port infrastructure investment projects do foster economic development and are important when a port is nearing its operational capacity. Under such circumstances, the lack of investments will clearly lead to additional externalities, namely congestion, which will undermine the competitiveness of a whole region, if not a nation.

THE ECONOMIC BENEFITS OF PORTS: DIRECT, INDIRECT AND INDUCED EFFECTS

Several economic impacts of port infrastructure investments obviously result into economic benefits. Economic theory often refers to ports as important factors of economic development, particularly from an historical standpoint where they promoted commerce and the welfare of nations. It is not surprising to realize that most of the world's major cities are port cities, even if in many cases port activity now plays a rather small role in the general economic framework of their regions. The basic argument is that ports expand the market opportunity of both national and international firms. By expanding the market areas of firms, ports increase competition, resulting in lower prices for the consumers of the port traffic. These involve all sectors of economic activity, including manufacturing firms, heavy industries, resource extraction industries or retailers. Therefore, the economic benefits of ports are specific to the nature of the hinterland they service. They can be straightforward for hinterlands heavily dependent on resources, since the output is directly handled by the port, or more nuanced when the hinterland is involving manufacturing firms producing intermediate goods.

Increasing competitiveness brought by port investments can also be a double-edged sword for a national economy. It enables foreign firms to better access a national economy and thus compete with national firms, with some sectors being put out of business. However, the benefits of having better access to foreign markets and cheaper goods usually far exceeds the risk of having inefficient national firms being undermined. At the aggregate level, increasing competitiveness promotes positive economic benefits, but these benefits are not uniformly distributed among sectors and geography.

Ports can be considered as "funnels" to economic development since they act as a catalyst and incite development to take place in specific economic sectors and locations nearby ports or along corridors. The economic benefits of ports are commonly categorized as direct, indirect and induced. Indirect and induced benefits are far from being clearly identifiable since it is difficult to demonstrate that the economic activity and use of the related resources would only occur as a result of the port investment. When port investment does lead to increased economic activity, the benefit is properly measured by the net value of the additional output. The direct benefits to the port are financial in nature and would be taken into account in any financial appraisal as well as in economic appraisals. However, the financial benefits would be valued somewhat differently, with economic appraisals using a social discount rate and for some inputs possibly valuing them at shadow prices.

ASSESSING THE ECONOMIC BENEFITS OF PORT INVESTMENTS

The economic benefits of ports are usually measured at an aggregate level by indicators such as value added, employment, taxation revenue and return on investment.

These indicators are primordial for the decision to invest in port development and must take into consideration:

- Demand forecasts trying to evaluate the expected traffic that the investment will support and facilitate.
- Liner shipping strategies, particularly how they service markets and how the port fits within their service configuration in terms of ship capacity and frequency. While some ports are acting as load centers, others are transshipment hubs. The function of transshipment is often the outcome of the strategy of a shipping company to service specific regions.
- Hinterland transport capacity and accessibility is contingent to the cargo that is bound to and originating from the port. It defines the existing and potential cargo base that could be handled by the port.
- Competition between terminals, since there may be competing terminals within the same port facility. Terminals in a monopolistic situation usually have more pricing power but can be linked with higher returns.
- Financing of investment relates to the capital source and conditions. Large port infrastructure projects are usually financed by bonds issued by port authorities or by investments made by international financial institutions such as development banks, sovereign wealth funds or pension funds.

Assessing this information can involve different methodologies:

- Surveys based on interviews and questionnaires or microeconomic data on firms. They try to identify and quantify the relationships between the various port actors, often from a qualitative perspective, but commonly in terms of employment. These studies have underlined the important relationships between freight forwarders and agents and that the economic benefits of ports are reflected in the complex system of transactions of the actors involved.
- Input-output models that seek to identify inter-sectorial multipliers, such as between port traffic and regional employment. They underline the agglomerating effects of port activities, either around the port or around the port region. Such studies have underlined low levels of elasticity between port traffic and service sector employment.

- Comparative analysis inferring economic benefits observed at a reference port, particularly its economic base. This approach tries to infer the economic changes that have already taken place in a comparable port setting (similar traffic and composition of traffic) to the port being investigated. Since local and regional economic conditions are not similar, such studies do not provide particularly accurate results. Still, they provide useful guidelines about what could happen to a port and its regional economy once an investment takes place.

Port activities have multiplying effects within an economy, which are much larger than the port itself. While the economic importance of port grows, particularly for the sectors they are connected to, their relative importance within the region they are servicing is often declining. There are thus diminishing total economic benefits for a regional economy as this economy grows and become more complex.

The following are the most commonly observed economic benefits of ports on regional employment:

- Port throughput is in general positively related to employment in port regions, implying that the higher the throughout the more employment. Employment impacts are more substantial in the industrial than in the service sector.
- Employment impacts vary by commodity sectors. Container and break bulk traffic have usually twice the employment impact than dry and liquid bulk traffic.
- Private ports usually have more regional employment impacts than public ports since they are usually servicing commercial supply chains.
- Each direct port employment is commonly associated with about 3 to 4 indirect jobs, although such figures vary widely according to the surveys and the context. There is limited empirical evidence about job multiplier figures.

However, the economic benefits per unit of port cargo handled, either in terms of employment or economic activity, usually increase with economic growth. The lower relative benefits of port investments are thus masked by economic growth, while in fact the economic importance of ports is increasing. The economic benefits are less directly related to port activities, but more related to the dynamics of the supply chains they support. This support becomes operational and functional, a benefit which is as crucial to national competitiveness. Therefore, global trends underline a decline of the direct economic benefits of ports, but a notable rise in their indirect and induced economic benefits. This trend is challenging because direct economic benefits can be readily assessed while indirect and induced effects are complex to capture.

GLOBAL-LOCAL MISMATCH OF THE ECONOMIC BENEFITS OF PORTS

With the setting of global supply and transport chains, there has been a growing level of mismatch between the benefits of port activities and the scale and scope of these benefits. While at the aggregate level it is clear that port investments have economic benefits, the spatial and sectorial distribution of these benefits is far less evident. One particular mismatch concerns local (community) versus regional / national / global benefits. The following points underline this mismatch:

- Labor usually comes from the local community and its benefits (mostly wages) are derived in the region, particularly indirect job multipliers. As port employment went down because of mechanization and containerization, so did the local labor benefits. Yet, several port jobs are remunerated at a wage which is much higher than those usually found in the manufacturing sector.
- Capital usually does not come from the community, but is either provided by national and international funding sources, such as investment banks, pension funds and terminal operators. The return on this capital (*e.g.* loan or operational revenues) thus does not accumulate in the port region but along global financial centers.
- Firms can be local in ownership but the commercial trends discussed above have underlinedvertical and horizontal integration in the port industry. This means for instance that terminal operation at one port is usually part of a portfolio of terminals located in different ports across the region or even the world. Thus, profits derived from terminal operations are not necessarily invested in the port they were generated.
- Port land use is usually regulated by leasing and concession contracts, but quite often land prices are a tool for attracting investors and they usually do not reflect real value. This underlines that the impacts of ports on real estate values are not necessarily fully accounted.
- Local transport infrastructure, namely roads, are usually provided for free (or at a fee lower than costs). This represents a form of local or regional subsidy for globally focused activities.
- Taxes and custom duties are just partly earned by the port region. They are usually a national source of income used to fund for other social and infrastructure programmes.
- Environmental (pollution) and social (noise, accidents) externalities are assumed by the community while the generators of these externalities usually bear only a fraction of them.

Port benefits are therefore increasingly distributed across actors and concern a geography that transcends the local community and at times the region. This trend skews the assessment of the benefits of port investments, where the local impacts can be much less significant than those at the regional or national levels. Conflicts or pressures can result from local communities that may be disappointed because their expectations about the economic benefits of port activities may not be met. Still, in spite of the complexity of assessing their economic impacts, ports remain fundamental to the economic well being of the nations, regions and localities they are embedded in.

4

Cruise Sector Policy in a Tourism Dependent Island Destination

CRUISE SHIPPING POLICY

The implementation of Cruise Shipping Policy has resulted in the growth of the cruise shipping sector in the country.

The details of the cruise ships handled in Indian Major Ports is as under:

Port	2012-2013	2013-2014	2014-2015
Chennai Port	5	3	6
Mumbai Port	41	37	47
Marmugoa Port	23	18	26
Cochin Port	41	34	40

The Government has taken the following steps to promote cruise and coastal tourism in the country:

i. Ministry of Tourism under its Scheme for Assistance to Central Agencies extends financial assistance to the Ports and Lighthouses for development of cruise related Tourism Infrastructure. The Assistance is extended for smooth operation of cruise liners that call on Indian Ports and also for up-gradation of existing cruise passenger terminals.

ii. Ministry of Tourism has identified theme based circuits under recently launched Swadesh Darshan Scheme to promote tourism in the country. One of the circuits identified is "Coastal Circuit" for development of coastal tourism infrastructure in the country, which will result in development of cruise tourism.

iii. To promote lighthouses as unique tourism destinations, some lighthouses have been identified for development by Ministry of Shipping. Out of these, two projects at Chennai and Mahabalipuram have been completed.

iv. Foreign flag vessels carrying passengers have been allowed to call at Indian ports for a period of 10 years with effect from 6th February, 2009 without obtaining a licence from Director General of Shipping.

CRUISE SHIP TOURISM INDUSTRY

FACTS AND FIGURES

There is a paucity of studies and little academic literature on cruise ship tourism despite it being the fastest growing sector in the tourism industry. With an 8 per cent annual growth since 1980, it has increased at almost twice the rate of tourism overall. A record 8.5 million people took cruises worldwide in 1997. The North American market (which includes the Caribbean) is the dominant one, and in 1997 it grew by 8.6 per cent to reach a record 5.05 million cruise passengers. In 1998, 71 cruise ships (which can carry over 93,000 passengers) from 24 lines plied the Caribbean, some year-round and some seasonally.

Although North America now accounts for almost 80 per cent of total cruise passengers, this dominance is expected to decline as other markets mature. International cruise revenue is estimated at US$17 billion a year. The Caribbean is likely to maintain its position as the most popular cruise destination in the world because of increasing preference for shorter cruises and an ever-younger market. The 2-5 day cruise accounts for some 37 per cent of the total product.

The Caribbean, accounting for 50 per cent of capacity in 1999, is the most popular cruise ship destination. Its convenient proximity to North America makes it an easily accessible 'pleasure periphery' for that market. Miami has ensured its place as the major hub from which most ships into the region operate, with up to 30 departures a week. Other major destinations include the Mediterranean (15 per cent), Alaska (8 per cent) trans-Panama Canal (6 per cent), west Mexico (5 per cent) and northern Europe (4 per cent). The length of the cruise season in these locations, however, is determined by climatic conditions. The South Pacific as a destination attracts only 2.2 per cent of the world's biggest and most lucrative cruise market, North America, and Australia's own cruise passenger generating capacity has remained consistently low and very specific in its product requirement.

The 'big three' cruise companies are Carnival, Royal Caribbean International and Princess that collectively control over two-thirds of the North American market. Star Cruises, a Malaysian-based company which caters primarily to Asian tourists aims to be the fourth largest. On the Asian ships a high percentage of passengers cruise in order to access gambling facilities that are not readily - or legally - available in their home countries, while the ships provide many activities to occupy their families.

TRENDS AND PERSPECTIVES

The mid 1970s was a period of international crisis in fuel supplies and costs; this affected itinerary planning for the Pacific considerably, with some shipping lines withdrawing from the market altogether or canceling cruises, both moves having long reaching effects on island economies. It was estimated that cruise passengers spent $40 a day in ports and the cancellation of two cruises in 1974 for instance, was estimated to mean loss of $100,000 to small businesses in various Pacific ports. P and O reduced the speed of its ships, which meant arriving at ports a little later and leaving earlier. The reduced port time affected the income of people like handicraft sellers and transport operators, the two sectors of Pacific communities which managed to benefit directly from tourism.

Numerous Caribbean islands now receive substantially more cruises than they do stopover tourists. Since stayover tourists are far more economically beneficial to Caribbean countries than cruise day visitors, the shift away from the former towards the latter represents an alarming trend for the region. Competition to add exotic ports of call has been intense, with Cunard adding 55 and Crystal Cruise Lines 24 new ports of call in 1998 alone.

The sharp increases in passenger capacity have been made possible by larger and still larger ships, whose economies of scale have produced record profits for the largest cruise lines. The size of many new ships, however, also dictates their cruising routes: they are simply too big to pass through the Panama Canal into the Pacific and are therefore restricted to the Caribbean and Mediterranean 'ponds'. They have flatter hulls than their predecessors and this, coupled with increasingly tall superstructures, means they are unsuitable for dealing with oceans and seas in those regions of the world that experience severe winds and currents *e.g.* the Pacific. (See very recent article from IMO News which reports on measures to ensure that the safety regime keeps pace with this trend towards ever larger vessels)

The cruise sector's ability to increase its passengers has been based on its success in reaching beyond its traditional upper and upper-middle class base into the middle-class mass market. Both the average age and the average income of cruise passengers have fallen steadily.

An important part of the strategy of the mass-market cruise companies has been to define land-based resorts such as Orlando and Las Vegas as their competition and to market their ships themselves as resort destinations. The ship is sold as the primary destination, not the ports it docks at. Indeed, 'destinational cruising' - where the ports are central to consumer choice and experience - is now considered within the sector to be a niche market.

The hotel and entertainment giants have been increasing their presence in the cruise business. It has been widely commented that the new mass-market ships seem more like floating theme parks, artificial islands largely replacing real-life destinations.

The rhetoric of globalization is very much evident in the pronouncements of cruise leaders. Cruise ships represent the ultimate in globalization: physically mobile; chunks of multinational capital; capable of being "repositioned" anywhere in the world at any time; crewed with labor migrants from up to 50 countries on a single ship; essentially unfettered by national or international regulations.

Globalization of the cruise sector has also led to increased internationalization of ownership and further concentration in this business, with a massive shakeout reducing the number of players. The pace of mergers, acquisitions, and bankruptcies has been dizzying over the past two decades. Globalization is also seen in the construction of a new terminal in the Grenadines. This will host mainly US-based cruise ships flying foreign flags and is being jointly financed by the European Investment Bank and the Kuwaiti Fund for Arab Economic Development, and constructed by a Kuwaiti firm.

The strong growth during the 1990s of the Asian cruise industry, particularly for Asian passengers, has interested industry watchers, especially in the light of the widespread Asian financial crisis of the late 1990s and the recurring political unrest in the region. Significant investment in infrastructure prompted by strong growth has ensured that Singapore will remain the hub for Asian cruise operators. Star Cruises has excelled in opening up the market but has also needed to invest substantially in appropriate regional infrastructure. They spent US$40 million at Port Klang, the port for Kuala Lumpur and US$12 million at Langkawi. Further developments are also planned for Phuket, Thailand.

The Japanese market remains elusive. Traditionally, the Japanese have preferred to cruise on Japanese-owned and operated ships and, unlike the growing preference for the cruise experience in other Asian markets, Japanese interest has remained more or less static since the mid 1990s, resting at about 200,000 passengers annually. In contrast, Koreans have taken to cruising enthusiastically, although they too prefer nationally owned and operated ships.

IMPACTS: BEST AND WORST CASE EXAMPLES

Cruise companies and the industry's professional organizations (CLIA and ICCL paint a rosy picture that all is well - see the Guide to Cruising from the ICCL web site. However, the following regional summary reveals a somewhat different and fairly consistent series of impacts and problems. But things are steadily improving, especially this year. For example, in June 2001 Alaska became the first US state to pass legislation regulating the industry and the ICCL announced that its members have unanimously adopted mandatory environmental standards for all their cruise ships.

Alaska

Several articles illustrate the heated debate in this State. The Alaska Dept of Environmental Conservation has been leading detailed discussions with the cruise ship industry, the state and federal government about waste management and disposal practices of cruise ships in state waters. This forum become known as the Alaska Cruise Ship Initiative and an Executive Steering Committee convened in January 2000 now directs the ACSI and its various work groups. Their web site provides full information about their efforts to improve environmental standards and take the cruise ship industry beyond minimal compliance.

The Bluewater Network campaign was obviously successful in lobbying lawmakers in Alaska to introduce bills that would require cruise ships to monitor and report regularly to the state environmental agency on all pollutants discharged into state waters and all wastes offloaded at ports.

Ted Manning cites the problems in Alaska where at some small ports up to six ships arrive at once in communities of a few hundred. He thinks 7 is the record in Seward. He is also anticipating a similar situation in Nunavut as more ships visit in the small window of opportunity in late summer. The largest settlement in Canada's north has less inhabitants than the average ship, and most only 100-300. A town like Pagnirtung with a beautiful fjord setting is beginning to attract several ships each year and there are fears of significant transformation of the local economy and society. This new territory does not yet have any strategy for dealing with the tourist influx.

Mexico and Central America

In 1998, 742 cruise vessels were reported in Cozumel and the maximum at one time was 11, although there are berths for up to 13 ships. With larger ships, many arriving at once, the numbers stress the existing capacity of the small town of San Miguel, as well as the available infrastructure. Last year Manning visited the San Blas islands of Panama where even their small ship (700) exceeded the population on the visited islands several fold -and at times two or more larger ships visit at the same time - totally transforming the port island into a circus of vendors.

Michael Lueck makes reference to a tv programme The Environmental Tourist - An Ecotourism Revolution which was produced by the National Audubon Society and TBS Productions in Washington, D.C. It has a section on an American diving cruise ship that goes to Belize and spits out about 300-350 divers at a time. He comments "this is an enormous amount and the worst thing is that the ship doesn't even dock in Belize. There is absolutely no benefit for the local people at all but the negative impacts on the fragile reef are tremendous." He didn't know if the ship is still operating because local people lobbied against it in a big way. (Belize Cruise Tourism Policy)

Bermuda

The executive summary of a report on the benefit of the cruise ship industry to Bermuda's economy poses some serious questions and it would be interesting to find out if their authorities have investigated them further.

Caribbean

Fantasy theming and simulation are endemic on most cruiseships. The ultimate in fantasyscapes on Caribbean cruises is not on the ship, however. It is to be found on "fantasy islands", privately owned by the cruise companies, off-limits to all but their passengers and employees, and marketed as the true Caribbean experience - only better. Of the 8 major cruiselines operating regularly in the Caribbean, six own private islands which they include among their ports of call. They are Half Moon Cay, Casaway Cay, Great Stirrup Cay, Princes Cay, Serena Cay, Coco Cay or at Labadee. The last is not actually an island but a piece of Haiti, surrounded by a ten-foot high iron wall, patrolled by armed guards. Disney dredged sand from the Casaway Cay bay and then ground it up further to make the island's beaches conform to a touristic image of Edenic perfection. It goes without saying that the development of private island destinations has been alarming to Caribbean countries, for in essence a local port is being cut out of the cruise itinerary in the process. The company reaps the economic rewards of renting their passengers everything from snorkelling equipment to cabanas to small boats, and selling them drinks and souvenirs at company-owned shops and markets. The already limited contribution of cruise passengers to local Caribbean economies is further eroded.

A further development of enclave-based encapsulation of cruise tourists (and their dollars) is the creation of private clubs for passengers in Caribbean ports of call. To some extent Caribbean destinations are imitating the cruise ships, introducing theming in the port city landscapes (such as Aruba, whose main street feels very much like a theme park) and creating manmade, artificial attractions, divorced from the geographical environment, as in St. Maarten.

The image of Dominica calling itself 'the Nature Island of the Caribbean' and free from mass tourism was severely dented in the late 1980s. It is the cruise-ship tourists (and their highly visible profile) who some critics believe have compromised the government's previous commitment to ecotourism. Cruise-ship arrivals rose spectacularly from 11,500 in 1986 to 124,765 in 1994. With a new cruise-ship berth at the Cabrits National Park, in the north of the island, and an improved deep-water facility near the capital, Roseau, up to 1,000 people per day pour off the cruise ships. Most of them take a whirlwind tour by minibus of a few of the island's best-known and most accessible sites. Ken Dill, a Dominican tour operator who specializes in hikes to the interior and nature tours warned that "if cruise ships develop to four or five a day and 4 or 5 times a week, it will be a turn-off for the ecotourists." Dill has to make sure his own customers do not bump into the cruise tourists when they visit the Emerald Pool, a short nature trail in the Morne Trois National Park leading to small waterfall. The carrying capacity of three sites, including the Emerald Pool, is being studied but there are some forestry experts who consider that places like the Emerald Pool will have to be 'sacrificed'.

In Grenada, the Tourist Board has called for stiffer penalties for crimes against tourists when one cruise line threatened to stop calling at St.George's because of passenger harassment by vendors. Pattullo writes that "it is the dumping of cruise ship waste that has been the focus of most concern. Cruises to the region have recorded a phenomenal growth rate and it has been estimated by the IMO that up to two kilograms of waste per person per day is generated. While some cruise ships have their own waste-processing facilities, many more do not." She makes the excellent point that "the attempt to clean up the ocean has also put extra strain on the land-based facilities of islands. In fact, the reason why not all countries have signed MARPOL is that to sign it would increase the pressure on their own land dumps. By not signing it, countries are not obliged to provide waste-disposal facilities and can refuse to accept garbage from cruise ships. Yet according to the IMO, which with the World Bank is organizing a project to deal with ship-generated wastes, this tempts cruise ships to dump at sea, whether legally or illegally."

Hawaii

A government article outlines the potential impacts and concludes the cruise market could represent a very valuable new visitor segment for their economy although there is the possibility of some diversion of visitors from land-based resorts. Taking advantage of this opportunity would mean making a commitment to improving quality of port facilities.

Tasmania

The potential problems of allowing cruise ships to visit the Port Davey/ Bathurst Harbour region of Tasmania's South West World Heritage Area are outlined in an article of Tasmanian Conservation Trust.

DISCUSSION REGARDING CONFLICTS AND BENEFITS

Ngaire Douglas in her chapter claims "the days when cruise ships routinely dumped waste overboard are over, although isolated instances are still reported. Large cruise lines such as P and O are keen to publicise their concern for the environment, while a number of smaller companies, whose reputations are built on their interaction with relatively remote and 'unspoiled' destinations, are taking an active part in the conservation of these destinations."

The Bluewater Network would disagree with this viewpoint. Along with their petition to the EPA they sent a report Cruising for Trouble: Stemming the Tide of Cruise Ship Pollution. Bluewater charged cruise ships produce enormous volumes of completely unregulated or inadequately regulated waste. Their executive director, Russell Long, said "It is time for the EPA to crack down on these floating cities that are having a severe impact on the environment." The success of practically the entire cruise sector is based on the use of flags of convenience (FOCs) to circumvent home country labor laws, taxes and maritime regulations. Labor laws protecting the rights of workers are virtually non-existent in FOC countries (most commonly Panama, Liberia or the Bahamas), and where they exist, they are often easily altered. For example, when it realized that Panamanian law guaranteed one day off each week, the cruise lines successfully lobbied for an exemption.

Most shipboard employees work seven days a week for six months at a time, with never more than a few hours off. Typically these employees are quite rigidly stratified in three groups: officers, staff and crew. On most vessels there is a clear ethnic cast to this hierarchy: Norwegian or Italian officers, Western European and North American staff and Asian, Caribbean and East European crew. Sexual harassment and rape are sufficiently common on cruiseships to have rated an article in the New York Times. (For further Information: Maritime Law) On all ships cruising the Caribbean, nationals from this region are a small minority of shipboard employees, no more than 7 per cent by one estimate, despite the proximity and high unemployment of the islands, a fact that remains surprisingly unexplored in the limited literature on cruise employment.

The economic impact cruise lines make to the Caribbean is matter of controversy (see summary in Patullo's chapter). She writes: "Conflicting statistics (from using different methodologies and multiplier effects), major leakages of spending, especially of duty-free goods, and a generally low contribution to the overall income generated by tourism in the Caribbean are themselves indicators of the economic limitations of the cruise industry. But even more fundamentally, who earns the money spent by the cruise industry? Who benefits from the government's expenditure on port and shopping facilities and such expenses as extra police security?" Moreover, "the extent of the interlocking of interests between cruise ships and local big business at the expense of local small business is at the heart of the debate about the cruise industry's economic contribution to the region."

Globalization detaches economic life from the constraints of geography - physical, cultural, political - and nowhere is this more evident than in Caribbean cruise tourism. The companies are entirely non-Caribbean. Their destinations are increasingly under their direct ownership and control; Caribbean cruises are taking on elements of 'cruises to nowhere'. The ships' laborforce is overwhelmingly non-Caribbean. What these ships do in the Caribbean Sea (including dumping) is outside the jurisdiction of Caribbean states. Meanwhile, the Caribbean Hotel Association complains helplessly about the unequal playing field it shares with the largely unregulated and untaxed cruiseships and worries about the stated ambitions of leaders to "empty out" its hotels. Indeed, the Caribbean Tourism Organization reports that the proportion of North American tourists who spend at least one night on land has declined from 61.8 per cent in

1987 to 48.6 per cent in 1998. Citing declining demand, American Airlines in 1998 forced Antigua, Grenada and St Lucia to pay it a subsidy of $1.5 million each in order to maintain daily nonstop jet service from Miami. With stopover passengers vastly outspending cruise visitors and therefore greatly preferred and sought after, smaller Caribbean countries find themselves having to subsidize their transport in order to get them.

Carnival Cruise Lines listed six aspects of the cruise 'product' which they said were superior to a land-based holiday: value for money; a 'trouble-free' environment; excellent food; the 'romance of the sea'; superior activities and entertainment; and 'an atmosphere of pampering service'. These factors are emphasized in cruise advertising, a constant presence on North American television and in the print media. If tens of thousands actively seek and wish to experience the above in 'pleasure prisons' then perhaps they should be encouraged to do so. By cruising instead of jetting off all round the world they are greatly reducing fuel consumption/emissions. It could be argued the trend towards larger vessels where the ship itself acts as a floating island is good for the environment as long as the most stringent onboard pollution and waste management controls are enforced. If these vessels do not dock at small islands then their potential impacts are contained.

EXAMPLES OF WORKING MODELS

The closest I came to identifying a working model where sustainable development and cruise ship tourism benefit from each other is in the Antarctic (incl. sub-Antarctic). The Arctic situation has evolved somewhat differently and policies to minimise/regulate the environmental impacts seem far less developed.

- *Antarctic tourism:* Island tourism with a difference by Dr Thomas Bauer which was published by our Islander magazine
- One on Antarctica that appeared in People and the Planet by Dr Bernard Stonehouse
- Sustainable tourism in the Arctic and Antarctic by Dr Bernard Stonehouse et al which was published in the INSULA journal
- New tourist threat to European High Arctic by Alan Small which again was published in Islander
- Environmental impact assessment of possible tourism at Marion Island prepared by the South African Dept of Environmental Affairs and Tourism in response to tour operators seeking permission to visit the Prince Edward Islands Special Nature Reserve
- The Falkland Islands Cruise Ship Industry by Debbie Summers of Falklands Conservation

In his article, Dr Bauer maintains that "unless very drastic changes occur (for example bigger ships) tourism to the Antarctic will continue to be a sustainable activity that brings great joy to those fortunate to have the opportunity to visit." This is due in large measure to the International Association of Antarctic Tour Operators which was founded in 1991. They have implemented guidelines or codes of conduct which appear to have been generally adhered to by their members.

Bauer's article states that "According to IAATO, the 1997/98 season, which lasted from Nov 97 to March 1998, saw 9,484 cruise passengers visit the Antarctic and the 1998/99 season recorded 10,027 passengers, the highest number ever in one season. The sixteen cruise ships and several yachts made 116 voyages." He has since told me there were nearly 15,000 passengers for the 1999/2000 season and the 2000/01 season figures should come in around the 13,000 mark. He also informed me that one recent development has been the emergence of some larger vessels like the Rotterdam but they fortunately don't land passengers or, if they do, land them at places they really can't do much damage.

Nevertheless, the trend in passenger numbers is upwards and the IAATO annual meeting held in June 2001 has a large number of items which shows they are actively addressing the potential problems and carrying out numerous research projects. Reference is also made to presentations on a South Georgia Island Future Tourism Development Plan and Development of Small Ship Tourism in the Falkland Islands. The latter refers to the study project undertaken by Falklands Conservation. Debbie Summers has since gone on to compile the Falklands Site Guide with Code of Practice for visitors.

Lindblad Expeditions is frequently quoted as a good example of working practice. Sven Olof Lindblad was among this year's winners of the UNEP Global 500 Award. Since 1979 his company have been providing small ship adventure cruises to destinations around the world and now caters for up to 12,000 passengers annually. His ships, when not ferrying tourists to exotic locations like Antarctica or the Galapagos Islands, are frequently deployed as floating conference centres bringing together environmentalists and policymakers to broker conservation agreements. The establishment of protected areas in the Bay Islands in the north coast of Honduras and in Mexico's Gulf of California was the result of one such meeting

Tours to the Galapagos Islands account for more than 20 per cent of Lindblad's overall business. The company conducts 5 per cent of all tourist trade to the islands. Since 1997 the company has experienced a 54 per cent increase in the number of guests travelling to the Galapagos. Through the creation of the Galapagos Conservation Fund, Lindblad has pioneered an approach that has raised more than US$500,000 since 1997, an average of US$4000 a week, in support of conservation projects in the Galapagos. Lindblad guests have contributed US$250,000 annually to the GCF. By comparison, the annual operating budget after salaries of the Galapagos National Park is approximately US$600,000. Lindblad has also set up a Travellers's Conservation Association, under the US Tour Operators Association, involving many companies which had previously had little involvement in ecotourism and conservation issues.

ANALYSIS OF THE MOST IMPORTANT FIELDS OF ACTION

A consistent point made is the lack of research and studies into the cruise ship industry. However, there would appear to be 20-30 academics around the world who specialise in this field. They cover different regions and bringing them all together to compare/share their findings and identify future priorities should be worthwhile if the LF decide to host a workshop. There is evidence that some are beginning to work on inter-regional projects and have a desire to form their own mailing list, discussion forum, create a web site, etc.

It strikes me a film or video examining the whole cruise phenomenon and its various impacts would be one of the best ways to highlight and graphically illustrate the contentious issues. The Television Trust for the Environment might well be interested and the person to contact is Zoe Stephenson. Bluewater's cruise ship campaign was featured in and contributed extensively to an in-depth tv documentary on cruise ship pollution hosted by the former President of the Sierra Club. This special episode of The Thin Green Line aired in late October 2000.

The economic impacts (costs and benefits) of the cruise ship tourism industry should be investigated and documented in a more consistent manner. In 1994 the Florida-Caribbean Cruise Association published a report carried out by accountants Price Waterhouse. This stated the average expenditure per person per cruise was estimated at US$539, which worked out at an average of US$154 per person per port. These figures were widely criticised. Interestingly, the ICCL Guide to Cruising makes reference to a recent Business Research and Economic Forecasting study conducted for the FCCA which found the typical cruise passenger spends almost US$90 at each port visited.

Falklands Conservation has expressed serious concerns about a Conservation Levy that was introduced by the Falkland Islands Government in 1999 under their Cruise Ship Ordinance. It is payable by all fare-paying passengers to the islands on vessels with more than 100 people on board. The charge is £10 per head and a total of £207,060 was collected last season. This season all vessels will be subject to the Conservation Levy which will increase the anticipated income to FIG by a further £10,000. This visitor tax is a misnomer because the funds raised are in fact not applied to conservation expenditure. Falklands Conservation feel that this additional levy money should be applied to expanding the conservation effort in the islands in areas such as increasing resources within the one man Environment Planning Dept, development of a National Biodiversity Action Plan and measures to protect the environment from visitor impact. Passengers and cruise ship companies are currently being misled into believing that such funds are being spent on conservation initiatives, and therefore feel no need to support these further (eg by donations/support to Falklands Conservation) and conservation work in the islands is being deprived of additional funding support. Other UK Overseas Territories in the Caribbean like the Turks and Caicos Islands which is threatened by a proposed huge cruise-liner port at the Ramsar site on East Caicos and the Cayman Islands where money from an Environment Protection Fund was transferred by Government to recurrent and statutory expenditure, indicates a strong need to address the issue of how cruise ship visitor taxes should be used to establish dedicated environmental funds for the UKOTs' in particular andSIDS in general. An investigation into why the different Arctic countries and the tour operators operating in this region have not formed a similar association to IAATO would be instructive especially as the number of expedition cruises to Nunavut, Greenland, Svalbard, Novaya Zemlya and Franz Josef Land are increasing.

The success of the Bluewater Network cruise ship campaigns (now Friends of the Earth> in Alaska and California might indicate that they could be widened to Canada in the first instance and then the Caribbean region where the North American market dominates and the environmental impacts are more acute.

Ted Manning suggested it might be possible to mount a study of risks associated with the cruise ship industry using the WTO indicators of sustainability approach for tourism. He is the leader of this programme and would welcome any approach. Prof. David Simmons is very interested researching into the energy/day, fuel/passenger kilometer type of figures. He has a suspicion that the pax/km energy will be much less than air travel and this may be an important consideration in the Pacific when faced with increasing concerns over climate change. Robert Wood states the study of cruise tourism represents a new opportunity, both for the understanding of tourism and its increasingly broad ramifications, and for the understanding of globalization in all its global/local complexities. Cruiseship crews are probably the most globally diverse yet physically compact labourforces anywhere. They constitute a virtual laboratory for studying what a truly global force might look like, and how global companies are responding to the challenge of both recruiting and managing such diverse aggregations of workers. They also represent a way of seeing how emergent global labor markets forge new connections and articulate with local communities and economies.

ECONOMIC RISK IN THE CRUISE SECTOR

1The cruise industry worldwide is subject to a wide range of risks, threats and vulnerabilities (collectively, risks). These risks can attach to any aspect of cruising and invariably, at some time, do. Risks can affect the cruise line itself, individual ships, ports and terminals, passengers, and onshore providers. The manifestations of these risks can be as high profile as building a new terminal which is unreachable by big cruise ships or more routine such as a powerful storm resulting in the cancellation of all outdoor shore excursions. The overriding risk for the destination, though, is that no ships will turn up because of a loss of confidence by the cruise line in that destination. 2Many risks are magnified in remote, isolated destinations as well as in destinations which experience rapid growth. New Zealand (NZ) falls into both categories: it is a two-day sail from any other destination (ie Australia and the South Pacific Islands), and as the following table illustrates, is experiencing rapid growth:

Table. Cruise Sector Growth in NZ.

Season	Cruises	Passengers	Total direct spend (NZ $)
1996/1997	27	19,400	$ 42.0 million
2009/2010	81	109,951	$271.2 million
2010/2011	96	138,242	$305.4 million
2011/2012	124	199,942	$347.3 million
2012/2013 (forecast)	130	210,000	$410.0 million

Therefore, it should be clear that destinations such as NZ can reap enormous benefits from the arrival of cruise ships. However, where mistakes occur, the cruise ships may stay away resulting in substantial economic loss to the destination. In other words, the stakes are extremely high.

Much of the discussion in this article focuses on the NZ cruise sector because of its vulnerability as a remote and fast growing cruise destination. However, the risks which are presented are no different from those faced by other destinations.

The primary difference to be borne in mind, though, is that any such risks are often exacerbated for destinations such as NZ where the loss of any scheduled port calls can have a more profound effect. In this article, the author explores the economic risks associated with the cruise ship-passenger-onshore relationship, evaluates the potential of these risks to threaten a destination's position as a competitive cruise destination and then proposes appropriate strategies which not only champion the development of a competitive cruise industry but also a sustainable one.

RISK, GENERALLY

Before proceeding any further, it is worth engaging in a brief discussion of the concept of risk. For the most part, risk is an indefinite term, with each definition providing a different interpretation. In its most common iteration, risk carries negative connotations (see eg Webster 1960 – "injury, damage or loss"). Roehl and Fesenmaier, on the other hand argue that risk is underpinned by "[a] choice [which] involves risk when the consequences associated with the decision are uncertain and some outcomes are more desirable than others" (1992, p 16). Thus, risk is no longer considered to be a matter of mere loss but instead involves choice. It is from this point that a discussion of risk in the cruise sector can commence. In other words, the risks identified in this article represent a loss of some degree but the strategies proposed in section 3 outline choices which may be available to manage those risks and therefore provide a foundation for a competitive and sustainable cruise sector.

A LIFE-CYCLE ANALYSIS

Whatever construct of risk is adopted, however, some logical structure or framework must be adopted in order to establish the context in which individual risks can be analysed and managed. On the basis of this requirement, the author takes the view that the adoption of a life-cycle framework is a pre-condition for the management of economic risks.

Owever, none of the extant tourism life-cycle models appears to be applicable to the cruise sector because of their emphasis on fixed tourism assets, eg a land-based resort or attraction (see eg Butler 1980; Hovinen 1981; Prideaux 2000).

Table. Cruise Sector Life-Cycle Model Based on the BA Model.

Life-cycle phase	Input
Product development Attracting passengers and ships *Tourist infrastructure – raw materials*	Itinerary planning – attracting ships Marketing and promotion – attracting (the right) passengers Leadership, coordination and knowledge – attracting local support
Infrastructure Operations, infrastructure and the regulatory environment *Infrastructure operations – production*	Air connections Ports and passenger facilities Regulation and the political environment
Distribution Moving, serving and assisting passengers onshore *Transport – distribution*	Transport Services Amenities
Use and consumption The onshore experience *Use and consumption - use*	Shore excursions Crisis and disaster
Disposal What the cruise passengers leave behind *Disposal*	Environment Economy

Note :

(a) Text initalics represents the life-cycle phases determined by BA (1994).

Instead, "cruise" involves tourism assets which, like the turtle, are at least partially carried around as part of the cruise experience during any given itinerary (ie mobile capital in the form of a cruise ship). Accordingly, the author expands the British Airways (BA) life-cycle methodology (BA 1994; Johnson 2002) to encompass and describe the full life-cycle encountered by cruise ships. In this article, the author uses the phases found in the BA methodology and adapts them to a cruise ship's transit:

APPROACH

8It should be evident from the discussion above that there is an urgent need to identify and manage the risks facing the high value, growing cruise sector. In simple supply and demand terms, the challenge for destinations is to ensure that the demand for cruising and its attendant risks do not outstrip that destination's ability to deliver a first class cruise experience.

The risks discussed in this article seek to address this concern, a concern which underpinned the author's discussions with cruise sector stakeholders both onshore and during two cruise voyages in and around NZ and Australia. These discussions produced a catalogue of risks which the informants felt could jeopardise NZ' success as a cruise destination. In addition to the interviews undertaken, the author also drew upon her own own experiences as an avid, frequent cruise passenger to provide a contextual background and critical view of the information gathered.

RISK – WHAT CAN GO WRONG BEFORE, DURING AND AFTER THE CRUISE

9In this section, many of the risks which face cruise destinations are discussed. However, this discussion should not be considered as a proposal to de-construct the cruise industry nor should any of the individual risks be considered in isolation. In fact, it is this very silo approach that has led to many of the weaknesses and vulnerabilities are perceived to exist within the NZ cruise sector and which undoubtedly also exist in other destinations.

PHASE 1: PRODUCT DEVELOPMENT

Itinerary Planning – Attracting Ships

It is the cruise industry version of a tautology that the development of a cruise sector by a port destination is dependent on that destination's ability to attract ships. Thus, the failure to do so will render a planned or developing cruise sector illusory, resulting in a loss of investment as well as revenues and jobs. However, the success of attracting ships is in part based on a complex itinerary planning exercise involving a series of factors which are not entirely within the control of the destination. Itinerary planning for NZ is governed by two inter-related seasonality issues: (1) the fact that NZ' own international cruising season is relatively short running from October until April; and (2) the practice of cruise lines to deploy their ships to the Southern Hemisphere during the Northern winter. Both of these factors, which effectively render NZ a "part-time" international cruise destination, mean that NZ is particularly vulnerable to itinerary-related risks. Some of those risks are set out in the following table:

Table. Itinerary-Related Risks in NZ.

Risk	Impact
Itinerary variations	The traditional 12 to 14 day Australia/NZ itinerary is being supplanted to some degree by voyages to the South Pacific islands, resulting in fewer visits to NZ ports
Rising fuel costs	Some NZ ports may be bypassed because of the expense of getting to NZ and the navigational challenges faced once there
Widening of the Panama Canal	A wider canal able to accommodate mega-liners will have a profound impact on itinerary planning, especially with respect to repositioning cruises because of increased flexibility to reposition ships between the East and West Coasts of North America, thereby potentially bypassing NZ

Marketing and Promotion – Attracting (the Right) Passengers

A passenger's decision to cruise is based on a complex matrix of many factors such as whether to go on a cruise in the first place, where to go, the choice of cruise line and the choice of ship (Petrick, Li and Park 2007, p 10). That decision can be motivated by such diverse factors as a desire to return to a familiar destination; own research; the influence of friends, family and travel agents; brand loyalty to a cruise line; or even preference for a particular ship (Petrick et al 2007, p 11). Cruise destinations must both understand and address these factors in order to attract not only cruise passengers but also the type of passenger who will make a positive economic contribution to the destination. Otherwise, the destination risks attracting passengers who spend very little whilst onshore.

In order to attract ships and their passengers, however, destinations must market to the cruise line executives who make the decisions as to where their ships will go. This effort can often be hampered by a lack of information and the consequent lack of understanding by the cruise lines of the destination. In cases where the destination's share of the global cruise market is small (eg

NZ), the cruise lines are unlikely to invest substantial sums to undertake their own research to familiarise themselves with those destinations. Therefore, it is incumbent upon the destination not only to market itself, but also to ensure that the information it provides is specifically targeted to those cruise passengers who it wishes to attract. However, these efforts must be counterbalanced against the reality that the cruise line's first priorities are to sell its ships and encourage passengers to spend money onboard, thereby potentially rendering the promotion of the destination superfluous to the cruise line's primary revenue-gathering objectives.

Leadership, Coordination and Knowledge – Attracting Local Support

The competitiveness, sustainability and therefore economic success of a destination's cruise sector can be threatened by a lack of leadership and structure. Evidence of this is apparent in NZ where port operations personnel have often acted on their own initiative to directly manage coordinating passengers' onshore activities, even beyond the port. According to two such individuals, their own proactive response is the result of the lack of an adequately funded national cruise organisation.

Other areas of concern include (a) the high turnover of staff within local or regional tourism organisations thereby making long-term planning difficult; and (b) the pre-disposition of national tourism agencies who are more attuned to the needs of tourists arriving by air rather than cruise passengers arriving and touring by sea.

Within the community itself, the community's response to welcoming cruise ships can make the difference between a positive port call (signaling the return of the ship on a regular basis) and a negative one (acting as a harbinger of elimination from the itinerary). Communities which exhibit negative or even complacent sentiments risk losing the economic benefits which would otherwise accrue to them.

PHASE 2: INFRASTRUCTURE

Air Connections

New Zealand's own small domestic cruise market and remote location make it highly dependent on air transport to supply its market. The vast majority of passengers who either join or leave their cruise in NZ (Auckland) must take an international flight of between three and 24 or more hours.

Therefore, fears brought about by terrorism, long flights and even the carbon footprints left behind can pose significant threats to cruising to remote destinations such as NZ (see Floyd, Gibson, Pennington-Gray and Thapa 2004, p 20; Roehl and Fesenmaier 1992, p 17; Sönmez, Apostolopolous and Tarlow 1999, p 15). Another significant concern is the availability of affordable connecting flights, particularly where the cruise embarkation city is far from major originating markets.

Expensive connections can result in a decision not to travel, as evidenced by the propensity of Americans to stay closer to home during the 2007-2010 recession (McKnight 2009).

PORTS AND PASSENGER FACILITIES

Terminal Location

The decision of where to locate a cruise terminal depends on a wide spectrum of factors including the ability of ships to navigate to the terminal; the terminal's proximity to air and surface transportation; and the type of port in issue. According to the captain of one ship, terminals in turnaround ports should not be sited in a congested city centre but instead, be built outside the city centre, along convenient transportation routes for air and ground transportation and city hotels. On the other hand, ports of call can and should be sited in or near the city centre to maximise passengers' time onshore instead of in shuttle buses or taxis. Other issues governing the physical siting of cruise terminals include:

Table. Considerations in Siting a Cruise Terminal.

Barrier	Barrier	Examples
Infrastructure barriers – natural or constructed	Low bridges preventing ships from reaching terminals	Sydney - Harbour Bridge; Brisbane - Gateway Bridge; Jacksonville - Dames Point Bridge; Shanghai - Yangpu Bridge
	Turning basins are too narrow for longer ships	Ships longer than 270 metres are too long to turn around in the Brisbane River
	Wharves are too short	Longer ships cannot dock at Circular Quay - Sydney
Neighbourhood attitudes	Pedestrian congestion resulting in local residents feeling overwhelmed	Charleston, South Carolina; Key West, small communities in Alaska
	Traffic congestion causing delays for passengers, supply vehicles and local residents	Circular Quay – Sydney, Princes Wharf – Auckland
	Noise emanating from ships' loudspeakers causing disquiet amongst nearby residents	Seattle and Brisbane
	Facility use leading to congestion and confusion	The entrance to the Princes Wharf (Auckland) cruise terminal is difficult to find because of its nestled position amongst shops, apartments and a hotel, meet and greet, baggage and marshalling areas for ground transportation are severely limited.

Terminal Facilities

Passengers' first impression of a port destination is often the port and its facilities. The port must be able to "provide a pleasing environment" (Marti 1990, p 161) and maintain excellence in all areas of service (Destination Victoria 2002, pp 50-51) because any negative experiences will have an equally negative impact on passengers' perceptions of the port and therefore the cruise lines' willingness for their ships to return.

Concerns about ports and their facilities are reflected in a quandary of Shakespearian proportions currently facing NZ, and that is: "to build or not to build," or more specifically, whether to build purpose-built cruise facilities, or not.

Ajamil (2005) notes that the demand for cruise terminals is predicted to rise significantly through 2017 but that the port industry has responded with low cost solutions, including utilising dockside warehouses. However, he further notes that cruise lines are beginning to expect terminals which are better located and have better amenities and facilities which can handle large numbers of passengers (Ajamil 2005).

With the possible exceptions of Princes Wharf in Auckland and Wellington's cruise terminal, all of NZ' cruise passenger facilities fail under Ajamil's definition of an acceptable terminal, thereby potentially leading to the sanction of a loss of cruise ship visits.

There is, however, the lingering issue of whether substantial investment makes sense. Jackson (nd, p 14) addresses this concern, stating that the level of investment in dedicated wharfs and cruise terminals is not justified "when compared to [sic] the level of return in visitor expenditure and revenue to governments."

In NZ, the decision to build will need to be balanced between (a) the port companies' willingness to invest in an activity (cruise) which they do not necessarily perceive to be either core or profitable business (TourismNZ 2008, p 12); and (b) the port community's willingness to provide welcoming facilities for cruise ships and their passengers.

An example of a port community whose significant investment did not result in the expected returns is the Canadian port of Nunavut. Port calls scheduled for the 2009 cruise season became erratic and were often cancelled (CBCNews.ca 2009).

Port Clashes and Mega-Ships

One of the most critical logistical problems facing many destinations, including NZ is overcrowding at its cruise berths, or port clashes. The challenges for NZ are not dissimilar to those facing the Caribbean where, according to Farquharson (2006, p 13), the key short-term challenge is to be able to accommodate the sheer growth in the cruise line industry and the parallel growth in the numbers of mega-ships. Atherley (2003, p 8) comments that "[a]t a minimum, a Port serving the cruise industry today needs at least two mega-ship berths if they are to make an impact." In the case of NZ the continued growth in cruising in NZ and its relatively short cruise season contribute to competition for its limited berth facilities.

In both cases, the desire to have more ships and more revenue needs to be balanced against the necessity to avoid port clashes because of the problems which can ensue.

The sheer size and capacity of mega ships also pose significant threats to the destination community itself:

Table. Threats Posed by Mega-Ships.

Locus of threat	Threat	Impact
Character of the cruise	Passengers may elect to remain on board – ship as resort	Passenger spend onshore will be less
Ports and port facilities	Length of wharf	Longer ships may not be able to dock or may interfere with port operations
	Passenger and crew exchanges	Insufficient hotel rooms and transport options to meet increased demand
Environment	More rubbish	Destinations may not be able to cope with the added burden and additional costs
Shore excursions	Increased demand	Local attractions may be forced to deal with long queues or close when mega-ships discharge their passengers into the port community
	Communities may feel overwhelmed	Restaurants, shops and roads may become over-crowded, causing dissatisfaction amongst local residents and land tourists

REGULATION AND THE POLITICAL ENVIRONMENT

Regulation

The operation of cruise ships within a destination in large part depends on government regulation which is consistent and fair to all stakeholders and which does not impose any extraordinary costs for compliance. In other words, regulation which is obscure, inconsistent or fragmented can pose a significant risk to the smooth operation of a cruise ship as well as lead to added costs for compliance. Some of the regulatory issues facing a destination's cruise sector include:

Table. Regulatory Issues.

Regulatory issue	Manifestation/Impact	Examples
Unclear, inconsistent or fragmented legislation	Difficulties in compliance	China: localised and frequently changing regulations make compliance difficult Australasia: no consistency as to whether and in which ports maintenance can take place
Lack of regulation	The cruise passenger experience becomes compromised	A lack of environmental regulation in some Pacific Islands has resulted in those ports becoming unacceptably polluted, compelling some passengers to remain onboard and not spend money onshore A failure to regulate taxis in Jamaica and Martinique resulted in cruise lines dropping those ports
Taxes, fees and other levies	Cruise lines may withdraw their ships from the destination	Charges can pose a significant threat to the economic viability of a cruise destination, especially where charges are high; ships may be withdrawn with a loss of tax revenues, jobs and direct spend revenues
Public understanding and compliance	Failure to communicate and explain	Passengers may not always have complete, relevant knowledge of a destination's quarantine regulations (eg NZ' prohibition on repatriating honey bought *en route* on shore)

The Political Environment

Public sector employees, including elected councilors can knowingly or unknowingly influence a destination's response to its cruise sector development. Public sector officials who understand the cruise sector can contribute not only to the welcome which cruise ships and their passengers are afforded, but more significantly, can also participate in the dialogue leading to a well-managed, commercially viable and highly rated cruise sector. On the other hand, public sector officials who have been described as obstreperous can act as barriers to the development of a sustainable and competitive cruise sector.

PHASE 3: DISTRIBUTION

Transport

Port and city shuttles can present logistical and financial issues for cruise lines and the service providers. The basis of provision varies and is a function of whether the cruise line funds the supply of shuttles, whether the passenger pays on a per ticket basis, or whether the destination provides them on a courtesy basis. However, with the provision of free shuttles comes the risk that the service may not be as frequent as the passengers would like. On the other hand, where passengers pay for their use of the shuttle, the expectation is that there will be sufficient capacity and frequency. One respondent commented that "either the passenger pays, or [it] costs the cruise company... if [the cruise lines] are charged, they will move out of those ports." Whether this is a view which is prevalent throughout the industry is not clear, but it was of concern to this senior officer and therefore should be considered as a potential risk.

Services

Retail: Shopping is a mainstay activity for cruise passengers (Dwyer and Forsyth 1998; Douglas and Douglas 2004). Many passengers will spend their entire port call shopping, whilst others will shop as part of their shore excursions or other activities.

Port destinations need to provide a welcoming retail environment which includes ensuring that the shops are open on those days when the ships are in port; that prices are not increased for the duration of port calls; and that all outlets provide passengers with a high level of customer service. A welcoming retail environment is particularly necessary in order to encourage passengers to spend money onshore in the face of the introduction of more shops and even malls onboard the ships (eg Oasis of the Seas).

Restaurants and cafés: The patronage of restaurants and cafés is an activity closely allied to that of shopping. Two issues are of particular significance where eating outlets wish to attract cruise passengers: (a) an understanding of passengers' eating habits; and (b) cost. Whilst there will be those passengers who seek to experience the local cuisine at any cost, the majority of passengers will seek out the best value for money, knowing that they can return to the ship for lunch if they find nothing suitable. In addition, destinations need to understand the national and cultural characteristics of visiting passengers. For example, Australasian passengers traditionally spend less on both onboard and onshore services than their American or European counterparts: meat pies are definitely "on" for Australians whilst the Europeans are more likely to head to upmarket cafés and restaurants with a respectable wine list.

Amenities

The provision of well-managed wharf side amenities such as trained "meet and greet" staff, informative signage, toilets, basic post office and foreign exchange services, and Internet and telephone services is one way to demonstrate a welcoming environment. A failure to provide foreign exchange services on all days when ships are in port can pose not only a source of major frustration for the passengers but can also result in substantially lower revenues for the destination because of passengers' inability to exchange their foreign currencies as and when funds are required.

PHASE 4: USE AND CONSUMPTION (THE ONSHORE EXPERIENCE)

Shore excursions

Cruise passengers represent a wide spectrum of interests, travel styles and expectations. However, there is one overriding benchmark for assessing both the successes and risks of shore excursions, and that is that passengers have a limited time in port. Overall, passengers seek port destinations which have "a good climate, access to an area possessing either a landmark of historical importance or an exotic or foreign culture" (Marti 1992, p 366). However, even in destinations where shore excursions rank highly (as they in NZ), there is the potential for a wide range of risks. These risks can attach to all aspects of the shore excursion planning and fulfilment process:

Table. Shore Excursion Risks.

Risk	Manifestation
Poor quality or an insufficient number of shore excursions	Cruise lines may withdraw ports because of an insufficient number of shore excursions or a lack of added value to those which were offered (see eg Northern Economics Inc 2002, pp 1-2 and 1-5)
Lack of flexibility	Passenger preference as to the content of shore excursions will vary - some passengers may wish to spend most of their time visiting one attraction whilst others will prefer more variety (Nilsson et al 2005, p 21)
Higher proportion of repeat cruisers seeking new experiences	Destinations must refresh their shore excursions offerings to continue to attract repeat cruisers
Changing tastes and increased sophistication of cruise passengers placing new demands on shore excursion content	Destinations must understand and be able to respond to changes in passenger interests, eg increasing popularity of culinary tours, visits to wineries and adventure activities
A wide demographic may render some shore excursion content unattractive or unsuitable	The increasing number of children and older passengers must be considered: (a) Children: (i) commentary on tour buses must be adapted to appeal to children as well as adults; (ii) tour bus operators should consider making booster seats available to enable children to look out the window; (iii) shore excursions may need to be added or modified to accommodate children (eg wineries and jet boating may not be appropriate) (b) Older and impaired passengers: shore excursion activities must be reviewed carefully to ensure that there are no barriers which could make it difficult for passengers to participate
Inability to manage the disparity between cruise tourists and land-based tourists	Because land-based attractions are for the most part designed for land-based tourists, the presence of cruise passengers over a relatively short period during any given day places intense pressure on those attractions, including "congestion, scheduling and control problems" (Atherley 2003, p 9)
The pre-disposition of destinations to focus on land-based tourism	A lack of understanding about and investment in cruise tourism can result in poor or inadequate service
High cost of shore excursions	Onshore visitor centres and attraction providers must be able to accommodate cruise passengers who opt to purchase their shore excursions onshore rather than purchase shore excursions onboard Onshore providers must be aware of and price their products according to the nationalities of the passengers: Australasian passengers are less likely to purchase the ship's shore excursions because they perceive them to be too costly

CRISIS AND DISASTER

Natural disasters, civil emergencies and operational risks can seriously impact cruise tourism thereby causing economic as well as personal damage or loss. Such events can affect both a decision to travel in the first place, render a port unable to handle ships and passengers and severely disrupt onshore activities:

Table. Risks Relating to Natural Disasters, Civil Emergencies and Operations

Risk category	Risk	Examples
Natural disasters and climate change	Earthquakes and tsunamis	Passengers can become stranded or trapped, thereby resulting in unanticipated direct and indirect costs to ports, passengers and ships
	Severe weather at sea	Ships may be unable to call into scheduled ports because of heavy seas or may suffer damage whilst docked
	Severe weather on land	Torrential rains can cause land and mudslides, trapping tourists and disrupting local transport
	Fires	Bush and forest fires can disrupt shore excursions
Civil emergencies	Piracy and terrorism	The actions of pirates and terrorists can cause injury, death or damage or delay journeys as evasive and defending actions are taken by the cruise ships
	Strikes	Strikes can disrupt port operations
	Military coups	Military coups can disrupt cruise ship visits
	Pandemics	Pandemics can result in the alteration of itineraries
Operations	Financial collapse	The financial collapse of transportation and other providers can impact cruise passengers both during their cruise and during transit to and from the cruise destination
	Mechanical failure of a ship	The mechanical failure of a ship can cause port calls or follow-on cruises to be cancelled

Any of the events described above can also lead to the cancellation of cruises. The effects of such cancellations are particularly acute for turnaround ports:

Table. Impacts of a Cancelled Cruise.

Resource	Impact
Ground transportation	Embarking passengers and crew who have already arrived in the embarkation city will be diverted to other facilities, thereby requiring transportation providers to respond
Accommodation providers	Passengers who have already arrived in the embarkation city may be forced to make significant changes to their travel plans placing unanticipated pressure on hotel rooms
Shore excursion providers	Although shore excursions in turnaround ports are generally limited to city tours as part of airport transfers, shore excursion providers will be affected
Air transportation	Passengers who have already arrived in the embarkation city will require changes to their air itineraries therefore placing demands on customer service support.
Port service suppliers	Contracts for provisioning and bunkering will be disrupted because of a non-existent or re-allocated need

In addition, there will be legal and financial ramifications for the port and the cruise line. For example, the cruise line will have contracted and paid for access to the berth several years ahead, whilst the local cruise industry may experience significant economic consequences resulting from a loss of visitor revenue and jobs.

PHASE 5: DISPOSAL

Environment

Shipping pollutes and pollutes absolutely. Although cruise ships represent less than 20 percent of the world's ocean-going fleet, they receive the most public attention for the waste they leave behind (Nilsson et al 2005, p 38; Brida and Aguirre 2008, p 3). Although the International Council of Cruise Lines (ICCL) has enacted a standard entitled the Cruise industry waste management practices and procedures and many jurisdictions have enacted their own legislation to regulate the disposal of cruise ship waste, where there is even theperception of harm (whether it is realised or not) that perception can pose a threat to port destination communities.

The Local Economy

Cruise tourism can also represent a threat to land-based tourism investment (see eg Wilkinson 1999; Atherley 2003, p 9). Negative impacts which can be experienced by port destination economies include:

Table. Negative Economic Impacts on Port Destination Economies.

Sector	Risk	Examples of affected destinations
Recreational sites and amenities	Commercialisation of local recreational sites and amenities previously used exclusively by residents	Key West; Juneau; Caribbean.
Local economy	Increase in seasonal, low wage jobs	Alaska
	Domination of gift shops and other tourism-oriented businesses which threaten or destroy local character; locally owned shops giving way to global franchises	Alaska; Key West
	Reliance on cruise passenger revenue, thereby making communities acutely susceptible to economic disruption from events such as changed itineraries and the withdrawal of ships	Alaska
	Displacement of traditional industries such as commercial fishing	Alaska; Jacksonville, Florida
	Introduction of shopping malls on ships displaces interest in shops and activities onshore, resulting in a loss of passenger-generated revenue into the destination	Global
	Flight away from land-based tourism uses, eg hotels being converted into condominiums resulting in the loss of tourism-related revenues and government taxes	Key West
Transport	Disruption to local access for cargo space and services as well as to local ground transportation	Caribbean
Fabric of the community	Disruption to the social fabric of the community	Key West

Thus, the overarching issue for port destination communities is whether a cruise tourism economy can be sustained whilst at the same time "minimising [its] undesirable consequences" to the community and its commercial sector (Hritz and Cecil 2008, p 176).

OPTIMISATION OF RISK – WHAT CAN GO RIGHT BEFORE, DURING AND AFTER THE CRUISE

Each of the risks identified in the preceding section can be managed through the application of appropriate countermeasures. However, it is not within the scope of this article to develop a detailed risk management plan for each of those risks. Instead, the author briefly outlines five key strategies upon which a framework for a competitive and sustainable cruise sector can be built. Each of the strategies outlined below has its foundations in best practice and as such can be considered to be acceptable tools for managing risk.

APPOINTMENT OF A NATIONAL COORDINATING COMMITTEE

The creation and empowerment of a lead body is key to building an efficient, competitive and sustainable cruise sector. A coordinated approach is necessary to address such diverse activities as marketing (McKenzie Wilson Network Partnership Group (MWP) 2004, p f); attracting cruise ships (Vojvodiæ 2003, p 205); creating a positive cruise culture (Farquharson 2006, p 18); mediating port clashes and overseeing port security planning (Vojvodiæ 2003, p 205); distributing funds received from a passenger tax for port projects (Smith 2009, unp); taking responsibility for development, training and knowledge management (MWP 2004, pp f-h); maximising the benefits of cruising to a region (Brida and Aguirre 2008, p 1); and securing the cooperation of stakeholders (Nillson et al 2005, p 18). A coordinating committee needs sufficient funding, an appropriate governance structure and sufficient resources to enable it to undertake ongoing industry research.

CRUISE CULTURE DEVELOPMENT

Concomitant with the establishment of a coordinating body is the promotion and introduction of a cruise culture. Destinations which seek to build a sustainable cruise sector need to recognise and appreciate the importance of cruise.

Development of a cruise culture requires a high degree of proactivity, stakeholder support spearheaded by strong leadership and the cooperation of three major constituencies: (a) the tourism sector as a whole; (b) the private (business) sector; and (c) the community itself (residents).

Tourism Sector

The successful creation of a cruise culture depends on the support of the tourism industry as a whole. One way to evidence this recognition is to include in the charters governing tourism organisations a statement outlining the organisation's understanding of the importance of the cruise industry.

Private sector

The second constituency is the private sector. In fact, the involvement of the private sector may be the primary catalyst for developing a competitive and sustainable cruise destination. The success of a destination may ultimately depend on the private sector's willingness to accommodate and cater to the needs of cruise passengers.

Community

The importance of securing the support of the community is vital. Destination Victoria (2002, p 54) argues that "[l]ocal communities likely to be directly affected by cruise-ship visits need to be actively embraced, informed and made to feel that any concerns have been addressed" whilst Jugovic (2007, p 709) argues that a port must work with the town to meet the needs of "ship operators, passengers, tourists, tour operators and others." Behnke (1999, p 24) echoes this view, advocating that the industry must be made part of the community through, eg inviting cruise line representatives to participate in any public consultations or local tourism advisory groups.

NATIONAL CRUISE MANUAL AND BEST PRACTICE

Destinations should consider the creation of a national cruise manual which documents all of the regulations, policies, procedures, codes of practice, customary practices and general advice relating to the operation of cruises within their waters.

A national cruise manual can provide a single source of accurate, up-to-date baseline information for all stakeholders, including the cruise lines and their ships.

It can also provide the basis for the development of a regime of continuous improvement as well as a mechanism for filtering out political biases and vested interests.

Underpinning the content of a national cruise manual should be a framework based on best practice.

Internationally recognised standards such as Total Quality Management (TQM) (Eraqi 2006, p 470) and ISO 9001:2008, Quality management systems (International Organization for Standardization 2011) can provide the basis for transforming the risks discussed earlier in this article into a competitive and sustainable direction for a destination's cruise sector. Adoption of a quality standard for the destination's broader cruise sector (as opposed to its constituent parts such as shore excursion providers, retailers or port management) can be considered to be a robust mitigation strategy.

EDUCATION AND TRAINING

The creation and implementation of a national cruise manual and the development of a cruise culture both demand and are dependent on a high level of education and training within the industry and the community. According to Destination Victoria (2002, p 51), training should be available to "service providers [including shore excursion providers] detailing cruise ship segmentation and implications of passenger spending levels and service expectations" (Destination Victoria 2002, p 51). Specific and substantial training about the cruise industry needs to be provided to all organisations and local communities which are involved with the cruise sector because cruise tourism is so fundamentally different from land-based tourism. In other words, it cannot be assumed that the rules, policies and practices which apply to land-based tourism apply equally to cruise passengers.

BUILDING CONFIDENCE AND RECOGNITION THROUGH BRANDING

Any given cruise destination may be "hot" for several seasons, but it will always be vulnerable to competition from other destinations. One of the most effective ways of communicating and building confidence and recognition in any destination is to develop a brand which clearly communicates its positive attributes and therefore establishes the product's brand equity. For example, whilst a significant amount of money has been spent developing a recognisable and highly successful brand for NZ tourism as a single destination (Morgan, Pritchard and Piggott 2002), no development has taken place of a brand or sub-brand which communicates the "hot-ness" of NZ as a cruise destination. Thus, as a strategy for mitigating the threat (risk) of competition, consideration should be given to the development of a cruise tourism brand.

CONCLUSIONS

At the start of her research, the author channeled her thoughts solely to assessing the risks which would ordinarily be associated with the movement of visitors through NZ ports, eg gangways blowing off their moorings, natural disasters, and the lack of dedicated cruise terminals and cruise berths. Whilst these and other "traditional" risks still form the core of the discussion found in this article, a significant, over-arching risk emerged and that is the economic risk represented by competition. Interviews with a broad range of stakeholders from many parts of the cruise industry gave rise to the premise that unless a destination such as NZ can more aggressively differentiate itself from other cruise destinations, it might find itself yielding its position to other remote or exotic destinations. Therefore, instead of focusing solely on risk as a negative concept, the author adopted the "choice" definition of risk proposed by Roehl and Fesenmaier (1992, p 16). In other words, this article focuses on conveying the message that there is an immediate need to find out who is most likely to cruise to a given destination; why they would consider a cruise to that destination; what will contribute to making their cruise visit memorable, safe and good value for money; and what they are likely to enjoy when they arrive. It is in this context that the author concludes that an overall, long-term cruise destination strategy is required which incorporates effective and appropriate marketing and promotion, infrastructure investment, product design, crisis planning, research and policy. It is a strategy which needs to be documented in a national cruise manual and managed, implemented and continuously improved upon by a national peak committee. It is the author's contention that only by managing all known and potential risks and promoting and building upon the destination's strengths can destinations such as NZ continue to evolve into competitive and sustainable cruise destinations.

Because many of the risks and impacts discussed in this paper are beyond the control of any single destination, further research needs to be undertaken as to how the cruise lines' global power affects an individual destination's ability to create its own competitive cruise destination. In the meantime, the challenge is to undertake a risk identification and analysis exercise such as the one proposed in this article so that the question posed at the outset can be answered by the cruise sector itself, ie whether the application of an accepted risk model can sufficiently optimise the economic risks which face a destination's cruise sector so that its growth can continue to be competitive and in the future, sustainable.

CRUISE TOURISM IN BELIZE

IDENTIFICATION

The Issue

Cruise tourism is the fastest growing sector of the Belize tourism industry. Its unprecedented and explosive growth, and its potential for acting as a catalyst for various support industries and related economic activities, has singled it out as one whose expansion and sustainability is highly desirable for the contribution it makes to the continued growth of the Belizean economy. [Graphic from Belize Tourism Board (BTB) web site]

As a relatively new industry, long term success and sustainability of the cruise tourism sector will largely depend on how the industry structure evolves to secure wide community participation and the consolidation of linkages with other economic activities (Jayawardena, 2002). It will also be affected by the extent to which the structure and stakeholder base of the industry provide for the dissemination of the benefits so that it results in socially visible economic progress.

The issue thus becomes: How can the economic benefits of the cruise tourism industry in Belize be made accessible to a wider cross section of Belizean society? In this vein, the objective of this case study is to conduct a critical analysis of Belize's Cruise tourism industry, with a view to identifying niche and support type opportunities for the integration of small and medium entrepreneurs and community organizations as active players into the tourism economy. Expansion of the stakeholder base would lead to a greater distribution of the economic benefits, a greater ownership and investment of the community in the product, and would advance attainment of the official policy objective "to assist development of tourism in Belize, optimize its contribution to the Belizean economy, while ensuring that growth is environmentally and socio-culturally sustainable."

It is important to note that the cruise tourism industry in Belize is regulated by the Belize Cruise Tourism Policy, one of the most sophisticated policy frameworks in the region, and that the recommendations contained in this paper recognize and respect the parameters established by said policy.

DESCRIPTION

A Growing Industry:

Cruise tourism is the newest and fastest growing sector of the Belize tourism industry. Cruise visitor arrivals grew from 14,183 in 1998, to 575,196 visitors in 2003, and one year later, in 2004, Belize welcomed 850,000 cruise visitors. New investments in the industry, such as the new US fifty million dollars cruise ship terminal that is being built in Belize City, promises to push the number of cruise visitors even higher in the near future.

In 2002, it was estimated that the tourism industry as a whole (including stay over arrivals), accounted for 15.6 per cent of GDP. Aggressive international marketing of Belize as a premiere "eco-cultural" destination also boasting marine activities, natural history and adventure type activities continues to translate into higher numbers of arrivals, both for stay-over visitors as well as for cruise visitors.

The Socio-Economic Expectations and the Political Context:

The explosive growth of Cruise tourism in Belize is a dream come true for tourism industry planners, but expectations surrounding its socio-economic impact must be addressed to prevent it from becoming a political and socio-economic problem that could threaten the industry and affect its long term desirability and sustainability. Of great concern is the perception that though cruise tourists disembark by the thousands, all they leave on shore is money in taxes and arrival fees charged by the government, and the garbage from their cruise ship prepared pack lunches. Many people, especially in Belize City, feel that they are being left without a fair share of benefits or opportunities from this apparently lucrative industry.

The rather simplified view expressed above is not far from the truth. There are few major onshore players directly benefiting from this industry, and these are largely the sizeable transport companies that operate tour buses and the vessels used to transport cruise passengers from the cruise ship to shore.

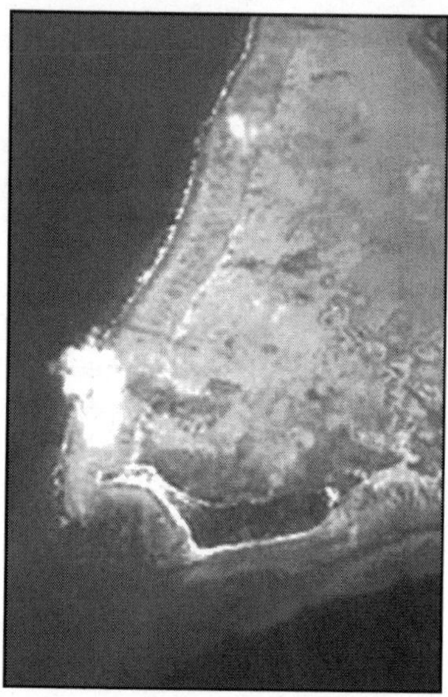

Fig. Half Moon Caye Natural Park.

Apart from these major players, only tour guides, drivers, taxi operators, the business owners in the tourism village where cruise passengers disembark on shore and a few other entities are seen to be receiving benefits.

The issue becomes more complicated because of its direct bearing on the issue of sustainability, given that a major part of the Belize tourism attractions fall into the c ategory of sensitive environmental areas, such as the archeological sites, the small offshore islands, the coral reef, the prehistoric cave systems and the rainforest canopy attractions. Environmental activists argue that without effective enforcement of the limits in the number of persons descending on these sensitive areas, high traffic to these places will lead to irreparable environmental damage, thereby killing the proverbial goose that lays the golden eggs. This perception that Belize's natural beauty and resources are being exploited and depleted by a few underscores the imperative of managing the expansion of the cruise industry by broadening the base of stakeholders and creating the conditions for more Belizeans to benefit.

Community Involvement and Ownership of the Iindustry:

Cruise tourism industry experts and industry writers agree that the long term success and sustainability of tourism (in the Caribbean) is predicated on its ability to ensure profitability, maintain product quality, secure community participation to strengthen linkages between tourism and other economic activities, and the eventual evolution into a diversified product such as eco, adventure, cultural or agri-based tourism. (Jayawardena, 2002). In Belize, the involvement of the average Belizean, the person in the street, is essential to ensure that these Belizeans feel invested in and develop a sense of ownership of this industry. It is only through a direct and beneficial involvement of small and medium entrepreneurs in the cruise tourism support industries that the perception of exclusion (of the ordinary person) from this industry will be removed.

The core cruise industry has tremendous potential to significantly create employment and generate economic activities through the creation of micro enterprises revolving around this industry. Locally crafted ceramic, glass and wooden souvenirs, music, cultural identifiers, and industry support services all have potential for development into economic activities that would give an average Belizean a stake in the industry.

Looking at Belize City and Beyond:

In terms of analyzing possibilities for employment generation and greater involvement, this paper will consider two main areas: (1) the regeneration of the inner parts of Belize City as part of the tourist package to bring economic activity to its residents (the new Cruise ship terminal is being built in Belize City).

What needs to take place is a quantitative analyses of the actual and projected numbers of cruise visitors, the estimated expenditure in the country, numbers of jobs created, and what impacts can be expected if a policy of expanding the stakeholder base is pursued; (2) The potential for increased economic activity in areas such as archeological sites, animal sanctuaries and other attractions outside of Belize City that form part of the tours and activities offered to cruise visitors.

This requires an analysis of the components of the typical tours offered to cruise visitors, to identify areas where the places visited and the services offered could be improved and expanded to encourage greater expenditure by the cruise tourist while on these tours. The employment and entrepreneurship opportunities generated through such efforts, *e.g.*: the establishment of visitors' facilities and souvenir shops near the archeological sites, would result in greater community participation and involvement.

It is also important to highlight that the analysis and any proposed recommendations will be formulated bearing in mind the imperative of ensuring synergies in policy, investments and infrastructure that will benefit both the cruise industry as well as the general tourism industry catering mostly to stay over tourists.

RELATED CASES

He concentration of this case study on two main areas for integration of additional stakeholders: (1) the regeneration of parts of inner Belize City, and (2) the expansion of services, products and tours / attractions, necessitate consultation with two types of case studies.

"The Cruise Industry and Port City Regeneration: the Case of Valleta," (Maccarthy, 2003) provides an excellent insight into the challenges, potential benefits, as well as negative developments to be guarded against when undertaking inner city regeneration for purposes of cruise tourism development. The main concepts that are explored in this case, such as the need to ensure that the community is involved, that infrastructure developments also benefit the community, that buildings and infrastructure investments have the best impact when they are laid out in mixed use areas, that the areas targeted for regeneration retain close economic and psychological linkages with the local community, that appropriate environmental, economic and social impact assessments are an integral part of any such development, all have a direct bearing and can be applied to the recommendation made in this case study for the regeneration of parts of Belize City.

The search for cases relevant to the second area of focus, the one pertaining to adding stakeholders by expanding services, increasing tours etc. has not met with much success. However, because the cruise tourism industry in Belize already operates within an excellent policy platform, focusing attention on expanding economic benefits and adding players is more a function of small business (product and services) development and poverty alleviation (training, job generation) than a strictly cruise tourism issue. This is the angle from which this second major area will be addressed.

AUTHOR AND DATE:

Nestor Mendez

Masters in International Policy and Practice candidate
Elliot School of International Affairs
George Washington University
Spring 2005

POLICY IMPACTS

The Belize Cruise Tourism Policy already provides an excellent legal and policy framework for the regulation of the Cruise tourism in Belize. It is clear about its sustainability objectives, assigns responsibilities, and points clearly at the goals of inter-sectoral economic linkages. This policy was designed and was agreed upon after consultations with all stakeholders and organizations. Moreover, it provides an excellent foundation upon which to cement the adjoining structures that will focus on delivering the developmental objectives such as job creation, the generation of foreign exchange, generation of income, employment and business opportunities (Belize Cruise Tourism Policy).

Social:

The potential social impact of implementing a full scale initiative to broaden the cruise tourism stakeholder base and to get the ordinary Belizean involved and invested in the industry could have tremendous social impacts.

Effect on Belize City – Regeneration of certain areas of inner Belize City to form part of the cruise tourism experience, and as a part of the wider tourism industry attractions could contribute greatly to addressing the social ills of the old Belize capital. The physical cleaning up, the infusion of money in the form of investments in small businesses, restaurants, bars, cafes, cinemas, could transform the "ghetto" like areas suffering social malaise of crime, violence, and unemployment into an economically active, safe and attractive part of the city. Effect outside Belize City – The social benefits to result from increased participation by local communities and businesses in the tours outside Belize City could also be positive and far-reaching. Empowering otherwise economically and socially secluded persons, such as women in rural areas who could set up vendor stalls in the villages en route to or near the tour destinations, could lift the pride and self-esteem of these communities and their people.

ENVIRONMENTAL:

Although the Belize Cruise Tourism Policy provides a sound basis for the sustainable management of cruise tourism, the potential for abuse will remain as long as the few inbound tour operators retain the monopoly and to a large degree the responsibility for enforcing established visitor number limits, and the official enforcement agencies remain deprived of adequate resources to do full scale monitoring. This lack of resources is more attributable to generalized economic conditions rather than absence of political will to furnish the necessary resources at the levels needed. In this light, expanding the base of stakeholders will result in a larger number of people being informed and educated as to the real economic reasons for protecting the proverbial goose, and their direct interest in the cruise visitor industry would mean that a larger number of watchful, zealous eyes will be protecting the environment. This can only result in better, more generalized efforts to protect the environment, and would certainly be an improvement over the situation that prevails today.

ECONOMIC:

It is anticipated and hoped that the major impact of an initiative to widen the base of stakeholders in the cruise tourism business will be in the economic sector. According to preliminary figures, Belize received approximately 851,436 cruise visitors in 2004. It is estimated that these visitors spent an average of

US $ 46.00 per person. Increasing the opportunities for the visitors to spend more money on shore is expected to result in an increase in average expenditure per person. Even an increase of US $ 20 per person would result in an infusion of more than US $ 17 million into the Belizean economy (over present levels), and with the number of arrivals expected to reach the 1 million visitors per annum range by 2006, increases in average expenditure per visitor will translate into even more significant cash inflows.

The extent of the economic impact of such inflows will require a formal study, but on the surface, the direct effects on employment, production and sale of locally produced goods will be easily visible, especially in the rural areas where there are less employment and economic opportunities. In Belize City, investment inflows for the infrastructure necessary for the regeneration of Belize City will have a tremendous impact on the city's economy, even before the direct economic benefits anticipated from the operation of the inner city as a tourist destination begin to be felt.

Fig. Afternoon in Belize City.

OTHER:

There are also other potentially beneficial effects that can result from the above initiatives. For instance, the University of Belize and the Centers for Employment Training will benefit from increased demand for trained workers for the industry. Beyond the usual expected training in services, bartenders, tour guides, craft producers, additional training will be needed in small business management, marketing, etc.

SUGGESTED INTERVENTIONS:

Two main areas will require policy interventions: The Belize Tourism Board, which is charged by the Ministry of Tourism with the management of the Tourism Industry, will need to intervene to help organize all cruise industry suppliers - inbound operators, service and goods providers, restaurants, community organizations and others stakeholders - into a functional operational structure. There will also be the need for the setting and implementation of standards of services and goods for provision to the cruise tourism sector. These standards will require wide consultation with all stakeholders, and its implementation will require the active participation of both the official agency (Belize Tourism Board) and the private sector.

LEGAL CLUSTERS

DISCOURSE AND STATUS/POLICY ISSUE:

The Government of Belize is very clear in its designation of the tourism industry of Belize as one of the pillars for the economic development and growth of the country. The legal and policy framework in place is excellent. The Cruise Tourism Policy is a clear manifestation of the recognition of this sector as an integral part of the Belize tourism industry, and one that the Government is intent on supporting and facilitating. In addition, since the expansion of the industry will of necessity incorporate merchandizing in certain cultural attributes of the country, it is important to note that the National Institute for Culture and History is very active in discharging its duties as the foremost promoter and preserver of Belizean culture. This institute has been recognized and accepted as an important partner in the development of the tourism industry, and it operates within an appropriately established legal framework.

FORUM AND SCOPE/EXISTING POLICY FRAMEWORK:

International – Components of the Cruise Tourism Policy of Belize incorporates the highest internationally accepted standards in environmental protection and best practices for sustainability. It is not expected that these standards will be relaxed. National - In addition to the strict environmental regulations in place in Belize, there also exist labor regulations to protect the rights of all workers. The existing policy framework pertaining to the three main areas covered by this case study: environmental protection and sustainability; preservation and promotion of culture; and labor related issues, appears adequate.

Regional - Matters dealing with the preservation of the environment, both marine and land based, are the focus of regional policy and cooperation efforts in the sub-region. The Belize Barrier Reef, the second longest Coral Reef in the World, is protected by an initiative known as the Mesoamerican Barrier Reef System Project (MBRS). The goal of this project is to enhance protection of the unique and vulnerable marine ecosystems comprising the MBRS, and to assist the countries of Mexico, Belize, Guatemala and Honduras to strengthen and coordinate regional policies, regulations, and institutional arrangements for the conservation and sustainable use of this global public good.

In terms of regional land-based resources, the Mesoamerican Biological Corridor initiative enjoys the support of Mexico, Belize, Guatemala, Honduras, and other Central American countries. Home to rare and endangered species as well as human inhabitants, the goal of the corridor is to integrate conservation, protection, and ecological balance within the framework of sustainable economic development.

The Mundo Maya Project, which seeks to promote the region as a single tourist destination, anchored on the ancient Maya traditional domains, is another regional effort providing a policy framework for sustainable cooperation in tourism and environmental protection. Funding for this project has been received from the Inter-American Development Bank, among other agencies

Local – Involvement of community level actors will need some organizing, although the legal and policy frameworks are already in place. In Belize City, the City council is empowered to take the lead in matters dealing with policy pertaining to tourism developments within city limits. Outside of Belize, in rural communities that can serve as gateways to the archeological sites, natural reserves and other attractions, are legally empowered to make decisions that affect their development.

In terms of the legal framework and facilitating environment for entrepreneurial activities, the Belize Trade and Investment Development Service is willing, able and legally capable to assisting.

DECISION BREADTH/STAKEHOLDERS/POLICY ACTORS:

The Cruise Tourism industry is tightly regulated, yet based on a solid partnership between the private and public sectors. The Belize Tourism Industry Association (BTIA) incorporates most of the tourism industry stakeholders, including subsidiary specialized sectoral organizations. Decisions that impact the industry are widely consulted, and it is expected that an increase in the stakeholders for the cruise tourism industry as previously described will be welcomed by the present players since the increase in services and attractions for the cruise tourists will also mean an increase in attractions for the overnight visitors, which are the principal target of the wider tourism industry.

The organization of specialized sector stakeholders into functional associations to promote and defend their interests is a welcome development and a sign of structural evolution in the industry. The Belize Cruise Tourism Association, the Belize National Tour Operators Association, and the Belize Tour Guides Association are some of the associations which are already formed and functioning, and which are working with the BTB and BTIA to make sure that the tourism industry policy in place serves their interests. It is anticipated that the number and type of these organizations will increase as the stakeholder base is expanded.

Indeed, it is imperative that community organizations and other potential stakeholders be facilitated with assistance to get organized as part of a strategy of expanding the stakeholder base.

LEGAL STANDING/LEGAL REGULATORY FRAMEWORK/ SUGGESTED POLICY INTERVENTION:

The Belize Cruise Tourism Policy assigns responsibility to the port agent for all matters relating to cruise ship operations in Belize, including licensing requirement. But beyond the operations of the ship itself, inbound tour operators also hold responsibility for the safety of the visitors. This means that an increase in attractions will need to take place in consonance with increased care to ensure that all new attractions adequately reflect certain safety and security standards.

It is important to note that the cruise tourism policy clearly articulates the safety requirements that are in place for all tours and attractions catering to cruise visitors and (overnight tourist). For instance, terrestrial tours require one tour guide for every fifteen tourists, and marine tours require one tour guide per every eight tourist. In addition, all sites included in tours clearly describe the type of activity, the level of exertion, age limits, and the type of health conditions that would prevent visitors from participating in the activities.

Belize's tropical climate will also demand that additional restaurants and other places where food is prepared and stored will need to be closely monitored for compliance with public health and safety standards.

TRADE CLUSTERS

Type of Measure:

Underscoring the thesis that expanding the stakeholder base is necessary to more widely disseminate the economic benefits, is the recognition that the stakeholders to be added are small and medium sized enterprises (SMEs), and community based organizations (CBOs) that will also need technical assistance with product development, business managerial skills, hospitality services, and familiarization with the policy frameworks in place. Equally critical will be the need to help them get organized and establish linkages with the inbound tour operators that supply the cruise ship industry.

The Belize cruise tourism product is a unique combination of nature based attractions, transportation, local food, cultural events and products, music, etc. Increasing the type and quality of services and attractions through delivery of the above mentioned assistance to SMEs and CBOs will give value added to the product, and will result in more expenditure on shore, and an increase in the number of persons directly and gainfully employed in the cruise tourism satellite industries.

An assessment of the impact of the expansion of the stakeholder base could be carried out by looking at two indicators:

1. The increase in the average expenditure per landed cruise visitor.
2. The increase in the number of persons gainfully employed in the cruise tourism satellite industries.

An Economic Impact Assessment would also be highly advisable in the early stages to measure the effects of the policies recommended by this case study.

RELATION OF TRADE MEASURE TO ENVIRONMENTAL/TOURISM IMPACT :

Directly Related to Product:

Negative: Widening the stakeholder base impacts the product directly, as it basically incorporates more actors in the assembly line. This increase in active players is also premised on the expectation of an increase in the number of cruise visitors per annum. Both increases, more workers and more visitors, have a direct bearing on the carrying capacities of the attractions, and thus potential for environmental impacts.

Positive: At the same time, an increase in the level of average expenditure per visitor and number of jobs created will result in product improvements in the form of additional investments and greater community interest. In addition, investments in economic activities and infrastructure that occur in response to the demand from the cruise tourism sector will impact positively on the total experience of the overnight tourists.

Indirectly Related to Product:

Enhanced image of Belize internationally as a haven for eco-tourists, and as an example of sound environmental protection.

Not Related to Product:

Poverty alleviation through employment generation as well as an improved standard of living in inner Belize City and in the rural areas involved in the tourism industry are the unrelated but welcome results.

Related to Process:

Growth in demand for indigenous products is contemplated as the visit (product) is enhanced. This will result in improved capacity for the production of these goods. Likewise, overall improvement in the delivery of services will also benefit overnight tourism and business visitors who may be tempted to return as tourists.

TRADE PRODUCT IDENTIFICATION/TRADE AND SERVICES:

The Belize tourism product is an experiential nature based / eco-cultural / adventure tourism product containing rainforest (eco-experience), coral reefs (diving), caving and rafting (adventure), Maya archeological sites and Mayan and Garifuna indigenous communities (cultural heritage).

ECONOMIC DATA

According to latest statistics released by the Belize Tourism Board, in 2004 the tourism industry accounted for 16 per cent of GDP, and employs approximately 25 per cent of the Belize Labor force.. Expenditures in the cruise tourism sector contributed about 14 per cent of the industry total.

The 2005 forecast for Belize by the World Travel and Tourism Council expects the travel and tourism industry in Belize to generate Bze $ 632 million of economic activity (total demand), with a direct impact of 6,450 jobs or 8.3 per cent of total employment, and Bze $ 171.2 million, equivalent to 7.8 per cent of total GDP.

However, since the travel and tourism industry touches on all sectors of the economy, in Belize the direct and indirect impact actually accounts for 15,422 jobs or 19.7 per cent of total employment, and Bze $ 438.6 million equivalent to 20 per cent of GDP. (World Travel and Tourism Council web site, 2005 country report for Belize, www.wttc.org)

With regards to the cruise tourism industry, an Economic Impact Assessment is necessary to determine the present and potential impact on the Belize economy. However, with the number of visitors expected to reach one million in 2006, any increase in the average expenditure will be multipliable by one million. Thus, an increase of even twenty US dollars per person would translate into twenty million US dollars (or forty million Belize dollars) which, given the size of Belize's economy, could have a great impact especially if the new economic actors are people who are newly employed.

IMPACT OF TRADE RESTRICTION:

External - The number of cruise tourism arrivals could be restricted by new international concerns for security while traveling. It could also be affected by severe natural disasters such as hurricanes that result in damage to the environment and the attractions. Internal - A restriction in the number of visitors below one million per annum would not be helpful for the cruise industry. Yet, setting a realistic limit is both desirable and necessary in the medium and long term in the interest of sustainability in keeping with established carrying capacity limits.

INDUSTRY SECTOR:

The international cruise line industry; international tourism involving cruise travel to Belize's port and nature and culture based attractions.

EXPORTERS AND IMPORTERS:

Exporters are the cruise companies, inbound tour operators, and all actors in the tour delivery. The importers are the consumers, the cruise visitors on whom the Belize Tourism Industry wishes to create such as profound impression that they return to visit as overnight visitors or at least as repeat cruise visitors.

MACRO/ENVIRONMENT CLUSTERS TOURISM POLICY CLUSTERS

ENVIRONMENTAL PROBLEM TYPE/ ENVIRONMENTAL ASPECTS:

The dangers of increasing the number of visitors and the number of attractions is that there will come a point beyond which it will not be possible to expand, and at which the sheer volume of visitors will start to have a detrimental effect on the environment, regardless of the monitoring in place. The extremely eco-sensitive attractions in Belize, such as the Coral Reef, the Rainforest Canopy Tours, the Natural Reserves, the underground cave systems, and the archeological sites all require a very careful monitoring and managing to make sure that the established carrying capacities are both realistic and implemented. The challenge after getting so many actors involved will be to agree on "reasonable" caps

RESOURCE IMPACT AND EFFECT:

Belize's main attractions are its natural resources, the rainforest, the archeological sites, the reef, etc. These are delicate and must be protected and preserved. They require a delicate balancing act between use and preservation, because their degradation and destruction would also destroy the industry.

URGENCY AND LIFETIME/URGENCY AND POLICY REVIEW:

It is urgent to get more stakeholders involved in the short term, but great care must be taken that their incorporation does not dilute the value of the product. The cruise tourism industry of Belize is such that it must be kept under constant evaluation to ensure that the resources are used wisely, and that additional investments are respectful of established standards for sustainability. The tourism industry of Belize is critical for Belize's long term economic well-being, and must be managed as such.

SUBSTITUTES/ALTERNATIVE POLICIES:

Short of the discovery of oil in Belize, or the development of a massive and dynamic information technology based industry that could provide employment and displace tourism as the most dynamic industry, there is no alternative to community involvement for ensuring sustainability of the cruise tourism industry and continued economic growth.

From the external point of view, the increasing number of cruise destinations in the Central American region has the potential for cutting into the Belize niche of pristine environment and a well managed experience. This means that because the product (the Belize visit) could be substituted by its neighboring destinations, the continuous enhancement of the product becomes even more critical.

OTHER FACTORS

CULTURE:

The issue of culture must be addressed from two perspectives:

1. *Preservation* – Given Belize's population of about 280,000 people, the majority of which are young, great care must be taken to ensure that the arrival of visitors totaling about four times the amount of the population does not erode the Belizean cultural identity through excessive exposure. In order to preserve the Belizean culture, cultural promotion must be an integral part of any enhancement of the visit experience.
2. *Promotion* – The Government of Belize's vision for long term tourism is the offering of a unique eco-cultural experience. In order to do this, culture must be conceived of as a national asset that requires nurturing for it to blossom and for it to be sufficiently strong to withstand the effect of massive merchandizing of cultural elements.

TRANS-BOUNDARY ISSUES:

Belize's main assets are heavily dependent on the natural environment. In recognition of this, Belize is an active participant in several regional initiatives such as the Mesoamerican Barrier Reef System project, the Mesoamerican Biological Corridor, and the Mundo Maya Project. From the perspective of protecting the natural resources, these initiatives have proven invaluable.

But since the Belize Tourism Industry has agreed with its neighbors on the need to promote the Central American region as a single destination for overnight tourism, then it will become important that experiences, standards and best practices be shared among the neighbors to ensure success of the initiative. These experiences will eventually permeate into the cruise tourism sector, and could potentially help in getting the countries of Central America organized such that as a group they could negotiate for better terms from the cruise lines that visit the region.

In the mid to long term perspective, there is a critical need for the Central American countries to begin considering a Central American Regional Cruise Tourism Policy which would encompass issues of environmental sustainability and enable them to better manage the cruise industry and regulate the cruise lines.

RIGHTS:

Belizeans have a right to benefit from an industry that is based on Belize's natural resources. They also have the right to have their natural resources protected and utilized in a sustainable way. The rights to have their culture respected and promoted are also balanced by the right of the cruise visitors to receive the best possible experience for the money they are paying. All these rights are not incompatible, and best reinforce each other when there is wide community participation.

RELEVANT LITERATURE

In the absence of significant specialized literature looking at the cruise tourism industry in Belize, an effort is being made to establish contact with and to hear the views of those industry players in the Belize Tourism Industry Association (private sector), the Belize Tourism Board, the Ministry of Tourism, the Tour Guide Association, etc. Literary sources dealing with Caribbean cruise tourism, with environmental protection, with regeneration of inner city areas, etc., are listed in the references section.

CONCLUSION

Policy Implications:

The most significant policy challenge facing this very successful industry is how to ensure that the benefits start reaching a wider sector of society. It is thus necessary to expand the stakeholder base, enhance the product, and grow the industry in a sustainable way.

More specifically, the Government of Belize will need to look closely at policy enhancements that will result in the setting of standards in the suppliers side of the cruise tourism industry. It will also need to assist these suppliers, especially entrepreneurs and community based organizations in getting organized to improve the product while benefiting from additional jobs and income. Education, training and institutional strengthening will be critical.

Recommendations:

- That the Belize City Council, along with the Belize Tourism Board and the Ministry of Tourism, commission a formal study that would include an economic impact assessment and social impact assessment, to determine the feasibility, cost / benefit analyses, and potential long term effects of encouraging a privately funded initiative to regenerate parts of inner Belize City to form part of the cruise ship and stay over tourism attraction. Such a study should also consider whether or not private entities would be interested in investing in such an initiative. [The product offered by San Juan Puerto Rico – the Old San Juan area – could be used as an example.]
- That all inbound tour operators be required, as part of the cruise visitor tours, to have at least one stop in a local community market, local restaurant, or local business establishment that would provide

a further opportunity for the cruise visitors to purchase Belize products.

- That the Tourism Training Unit be strengthened and deployed as the primary agent to work with the Village Councils Associations to facilitate the organization of community based organizations as suppliers of products and thus become cruise industry stakeholders. The initial villages to be targeted should be those communities with proximity to or on the way to established and potential cruise visitor attractions.
- The BTB and the BTIA need to work together to assist the establishment of linkages between the tour operators and small and medium sized enterprises, and community based organizations, so that these new stakeholders become product and service suppliers for the cruise industry.
- That the Belize Tourism Board undertake a publicity campaign through radio and television in Belize to :
  - Highlight how important the cruise ship industry is to Belize
  - Show how everyone has a part to play.
  - To talk about what benefits can be derived by getting involved in the industry
  - To highlight employment and entrepreneur opportunities that exist.
- That the University of Belize design and offer sixth form level courses in ceramics and glass making that would provide students with the technical know-how to produce souvenirs for the tourism industry. Once these types of production facilities are established, they can also be used to produce plates, ash trays, etc for the hotels and restaurants.
- That the National Institute of Culture and History strengthen its efforts to encourage and support local artists, and organize workshops for musicians and singers to let them know how to go about contacting the local recording studios for the production of their music.
- That the National Institute of Culture and History and the Belize Tourism Board work together to make sure that local Belizean music is played in the buses, in the restaurants, and in any other places visited by cruise and overnight tourist (where music would either be played as a matter of course, or where it would be a welcome addition.) Furthermore, locally recorded CDs containing the music being played should be available for sale where it is being played, including all souvenir shops.
- That the Ministry of Agriculture, through its countrywide network, encourage the production and preparation of fresh fruit juices and

other local products such as coconuts, Belizean sea weed drinks, etc. to be available for sale near the attractions visited by tourists.

- That the Belize Tourism Board be pro-active in working with all sectors of the tourism industry in the joint establishment of standards of service that could be evaluated and promoted.
- That the Ministry of Tourism promote a policy of assisting entrepreneurs through the provision of basic business and tourism relations training, to ease their emergence as active players in this industry. That the Tourism Training Unit of the BTB be strengthened and institutionally linked for better coordination with the University of Belize and the Centers for Employment Training.
- That the Belize Tourist Board (Gov.) and the Belize Tourism Industry Association (private sector) team up with the Belize Credit Union League and other such financial institutions, to secure low cost financing for new businesses and expansion of existing ones.
- That the Ministry of Tourism work closely with the Centers for Employment Training to ensure that the training programmes being offered are reflective of the tourism labor market demands.
- That the Ministry of Tourism, the Belize Tourism Board, the Belize Tourism Industry Association, the Belize Cruise Tourism Industry Association, the Belize Tour Operators Association, and other industry stakeholder, together request the Cruise Lines to partner with them for the establishment of an eco-cultural trust fund to be financed with voluntary contributions by cruise visitors. Funds collected would be invested in small conservation and cultural promotion projects in Belize determined by a trust oversight committee set up specifically for this.
- The voluntary contributions by the visitors could entitle them to in kind returns (dinners on board the cruise ship, shows, or on shore to dinners, tours etc) valuing up to 50 per cent of their contributions. These in –kind returns would be provided by the cruise line and the on shore industry stakeholders. Such a mechanism is being successfully employed by a cruise line dedicated to taking ecological tours to the Galapagos Islands.
- That the University of Belize, along with the Department of the Environment and other participants as deemed appropriate, be requested to undertake a study to determine the feasibility and viability of recycling used glass and aluminum material into souvenirs and other usable goods. A pilot project could initially focus on recycling Belize city refuse, and if it proves viable, this could expanded to recycling presorted used glass and aluminum refuse from the cruise ships, to be transformed into souvenirs and sold to cruise visitors.

The environmental, earth friendly, and publicity potential for such an industry could be tremendous.

- Finally, this paper joins in supporting an expeditious elaboration of a National Tourism Policy as announced by the Hon. Godfrey Smith, Minister of Tourism on April 6, 2005. The initiative to implement a National Artisans Seal and Seal of Excellence for the promotion of made in Belize products are also highly commendable and should be concretized as soon as possible. Both of these major policy initiatives are fully congruent with the other recommendations posited in this paper.

5

Cruise Ship Tourism and the Environment

CRUISE SHIPS AND THE ENVIRONMENT: THE STATE OF THE SITUATION

The cruise industry is a sector of tourism with two characteristics that are important to understanding its environmental impact: its rate of growth and spatial concentration in the Caribbean Sea. This rapid rate of growth combined with the tendency to concentrate on or near ecologically dynamic coastal environments (many of which are considered biodiversity hotspots) in the Caribbean raise important policy questions about how to preserve the natural resources upon which the industry depends.

In terms of measuring the size and growth of the cruise industry, different sources use different indicators, including number of passengers, number of beds, number of ships, etc. but all agree that cruise tourism is one major growth area of the travel industry. By one measure, cruise tourism as a global phenomenon has demonstrated an 8 per cent annual growth rate since 1980 (Johnson, 2002) which is around double the growth rate of tourism overall. A cruise industry overview notes that 22 new state-of-the-art ships were contracted to be added to the North American fleet in 2009 (FCCA, 2009). The newest cruise ships are bigger and bigger. They rely on cutting edge technological features as well as economies of scale. In some destinations and ports of call, issues of congestion have emerged, and the environmental carrying capacity of the entire Caribbean region is being questioned.

The other striking characteristic of cruise tourism is its market concentration, spatially. The markets are highly concentrated around U.S. passengers in U.S. or Caribbean waters. In fact, six of the world's eight leading cruise markets are in or adjacent to U.S. waters (National Research Council, 1995, p. 47). The dominant world destination is the Caribbean region, which accounts for 50 per cent of total world market share. Insiders are quick to point out that the distribution of industry benefits is not concentrated in a parallel way. "Sixty-five percent of the cruise industry's profit comes from the Caribbean, but only 7 percent of their employees come from the Caribbean, and only 1 percent of the taxes they pay come to the region," reported Allan Chastanet, former Board of Directors for St. Lucia to Fortune Magazine, 1999. So, there is clearly a great need for the Caribbean region to capture more economic "rent" on the public goods it is providing to the industry.

Various information sources about the environmental implications of the cruises list a variety of different impacts that cruises have on the environment. An LCA (life cycle analysis) methodology identifies five areas of environmental impact: infrastructure impacts such as ship construction, operational impacts such as energy use and anchor damage, distribution impacts such as the transferring of people to and from departure points, use impacts such as congestion and water consumption, and waste impacts such as sewage, plastics and hazardous substances (Johnson, 2002). The most commonly mentioned problems are the generation of waste, and anchor damage to coral reefs.

Solid wastes. In the Caribbean, the links between tourism and solid waste are clear. Tourists in the Caribbean have been estimated to generate twice as much solid waste per capita as local residents (World Bank, 2001). The solid waste generated by cruise ships can be measured in several ways: by boat per trip, by boat per day, by passenger per day, etc. According to Davies and Cahill, a cruise ship carrying 2,700 passengers can generate at least a ton of garbage per day. An average passenger generates 2 pounds of dry garbage and one half pound of food waste.

Hazardous wastes. Cruise ships produce toxic chemicals and hazardous waste from dry-cleaning procedures, used batteries, and paint waste from brush cleaning (Malbin, 1999). On-board photo printing and processing also creates toxic waste, as well as sludge-contaminated filter materials. Sewage and grey water also present public health hazards if not properly treated. A typical cruise ship can generate 1,000,000 gallons of grey water on a one-week voyage (Johnson, 2002). There is evidence that disposal of wastewater and untreated sewage into the sea has already curtailed tourism activity in some areas, such as Negril, Jamaica, where there is less diving tourism than before (World Bank,

2001). Anchor damage to coral reefs. The coral reefs are some of the most delicate and ecologically valuable ecosystems of the world, and are also at great risk from tourism activity. Much damage is incurred from snorkeling and diving, but there are also reports of anchor damage to coral reefs and sea floors caused by cruise ships. For example, In October 1988, a cruise ship dropped its anchor on a coral reef in Virgin Islands National Park, St. John, creating a distinct scar roughly 128 m long and 3 m wide from a depth of 22 m to a depth of 6 m. The anchor pulverized coral colonies and smashed part of the reef framework (Rogers and Garrison, 2001).

EXISTING POLICY TOOLS AND THEIR SHORTCOMINGS

The major environmental concerns for the cruise ship industry are congestion and limited environmental carrying capacity due to its rate of growth combined with its waste generation and impacts on fragile natural ecosystems such as coral reefs. The current regulatory framework is insufficient in addressing these concerns. Regulatory tools can be categorized into three major kinds: normative-based or command-and-control legislation, price-based or incentive-based tools, and voluntary self-regulations. All three kinds of regulation already exist for the cruise industry, but they are not unified or completely effective in addressing the environmental problems that the industry imposes.

Command-and-control regulation of the cruise ship industry is characterized by a lack of regional policy and a lack of local enforcement, since actors are spread over several countries. The current international regulatory body is MARPOL 73/78, the International Convention for the Prevention of Marine Pollution, which has established standards and guidelines that all sea vessels, including cruise ships, must meet. However, these regulations were deemed insufficient in protecting sensitive, high-traffic areas, which led to the Wider Caribbean designation as a "special area" in need of further regulation in 1991 (World Bank, 2001). MARPOL guidelines for dealing with ship-generated waste are also insufficient because, like many command-and-control regulatory measures, Caribbean governments lack the resources for proper monitoring and implementation. In the Caribbean the number of prosecutions has been less than in the United States, and it is not anticipated that the enforcement capabilities of the Caribbean will dramatically increase. Similarly, the U.S. Coast Guard has "Operation Cruise Watch" initiatives in Alaska, Florida, and California, but there are no indications that less developed nations are prepared to apply the same level of control (Johnson, 2002). Governments of the less-developed world have far greater priorities, *e.g.* economic development, health, welfare, and education, for their limited financial resources than environmental monitoring (Hall, 2001). Regulatory command-and-control instruments fall short in addressing environmental problems.

Regulations involving market-based incentives for cruise ships have been minimal and scattered. These policy tools generally take several forms, including taxes and subsidies. An example of a subsidy that is already in place is A Global Environment Trust Fund grant of US$5.5 million. This grant is designed to provide technical and legal assistance to the Caribbean region's countries in implementing MARPOL 73/78. The subsidy highlights the need for a coordinated, regional effort in policy design and enforcement (World Bank, 2001). When evaluating a subsidy as a form of price-based incentive, it is important to criticize that it does not meet the "polluter pays principle," meaning there is no penalty imposed on the source of the problem.

In terms of tax-based market incentives for cruise ships, tax legislation has been difficult to apply due to the political-economic structure of the industry. In addition to being geographically concentrated, the industry is also heavily economically concentrated, with a few major corporations representing the vast majority of the market. The top three parent companies, Carnival, Royal Caribbean, and Star Cruises, account for 91.5 per cent of the North American cruise market share, making the cruise industry one of the most concentrated in the world.

These major cruise lines are registered in countries such as Panama and Liberia outside of the U.S. in order to avoid various tax, labor, health, safety, and environmental regulations. Indeed, the huge lobbying power of these corporations has granted them substantial tax breaks and other savings that contribute to the competitiveness of cruise prices (Lumsdon and Page, 2004). In other words, corporate cruise line giants depend on lobbying power and deterritorialization in order to offer a competitive product. One port authority observes, "some sub-regional countries have tried to introduce an environmental levy to compensate for the costs they incur in dealing with solid wastes that are discharged by passengers while ashore. Needless to say, this has been met with the usual threats and many countries have had to absorb this cost by granting further concessions to the Lines" (Atherley, 2003).

Despite the political-economic clout of these major corporations, a few individual Caribbean countries have asserted their right to impose regulations and incentive-based policies on the cruise industry. In 1996, six small countries in the Eastern Caribbean (Antigua, Dominica, Grenada, St. Kitts, St. Lucia, and St. Vincent) jointly decided to charge a US$1.50 per passenger "cruise ship waste disposal fee" to finance environmental clean-up and conservation (The Nature Conservancy, 2004). Bermuda, as the strongest example of both normative-based and incentive-based policy, limits the number of regular callers to six vessels per week, 6500 maximum per day and an annual maximum of 200,000 visitors. Bermuda also has the highest cruise ship passenger fee in the region, at US$60 plus $20 for overnight in high season and $15 during the off-season (Limited, 2005). The Bahamas have set a head tax at US$15 (Johnson,

2002). Arguments have been made for further implementation of capturing economic "rent" on natural resources through taxes and fees, following examples such as Berumda and the Bahamas.

Finally, voluntary self-regulation is probably the most common environmental policy tool in place for the cruise ship industry today. Initiatives from within the industry have been used to address environmental problems and to pre-empt further top-down regulation. Individual corporations project themselves as environmentally responsible through campaigns such as Royal Caribbean's "Save the Waves" and Holland America's "Seagoing Environmental Awareness" campaign, although the argument has been made that these are primarily public relations tools (Klein, 2002).

Volunteer associations, non-governmental organizations, and certification bodies offer a more unified effort by the industry to impose environmental standards upon itself (Hall, 2001). Examples include Green Globe 21 certification for various resorts in Barbados and Jamaica, and a current project to adapt the Blue Flag certification of of water quality programme from Europe to the Caribbean region (World Bank, 2001).

POLICY PROPOSAL: ARRIVAL FEES AS AN ENVIRONMENTAL TAX ON CRUISE SHIP TOURISM

Environmental taxes on tourism Is an area of tourism policy that deserves attention, due to the dual function of environmental protection and revenue generation that such taxes can provide to ecologically fragile, less-developed destinations that are in great need of both. This is often referred to as a "double dividend." The reasons for taxing tourism are "linked to its capacity to act as a substitute of price for the public goods and services consumed by tourists" as well as the "corrective (*i.e.* environmental) role that these taxes can be given." So the three justifications of a tourism tax are often to generate revenue, cover conventional costs of public services, and internalize external costs, such as environmental ones (Gago, et al).

Environmental and tourism taxation has taken on many forms. Tourism is taxed both indirectly by taxing tourism business and by taxing the tourist directly through consumption taxes and special tourism taxes such as entry and exit fees. Environmental taxes in tourism are ones that are designed to internalize external environmental costs that are not reflected in market prices. In order to be considered an environmental tax, a tax must meet the design criteria of internalizing external environmental costs and not simply the criteria of generating revenue that may be earmarked for environmental projects. The tax base must be the physical unit that has a proven specific negative impact on the environment. The idea of an environmental tax, then, is to correct the market failure of too-low prices that create a gap between an inefficient private optimum and an efficient or optimum social solution (Palmer, Riera, 2002).

In the case of cruise tourism in the Caribbean, the need for further environmental taxation on tourism is clear. Caribbean countries are failing to capture taxation revenue in proportion to the services they offer. It is not collecting sufficient "rent" on the infrastructure services, public goods, and natural resources it is providing. The result is a market failure and inefficient allocation of resources.

As World Bank points out, rents are a type of payment for the use of a resource, and in tourism these natural resource rents often go un-captured. Therefore, an increase of user fees would help to capture rents at the source: the tourist and the consumer surplus derived from the use of the natural resource. A case is made for a direct tourism tax: "although tourism operators should be expected to pay their fair share of taxes like any other good corporate citizen, taxing tourists is likely to be the most effective means of rent capture." (World Bank, 2001).

A good policy proposal for the environmental problem of cruise tourism in the Caribbean would be arrival taxes for cruise ship passengers. This tool is already in place in Bermuda and the Bahamas, but all Caribbean nations with cruise ports of call could highly benefit from such a tax. The World Bank includes this as a policy option, but specifies that the arrival fee should be explicitly identified as environmental or resource user fees, so as to be more acceptable.

IMPLEMENTATION, EVALUATION, AND PREDICTION OF CONSEQUENCES.

The implementation of an arrival fee per capita for cruise ship passengers at each port of call would mirror the systems that are already in place in destinations such as Bermuda and the Bahamas. However, due to the political problem of the balance of power against small Caribbean countries and in favour of large corporate interests, a better solution may be a region-wide policy of arrival fees implemented through a voluntary association of Caribbean governments. This could help governments gain leverage when imposing a tax on the cruise industry. This could also prevent loss of competitiveness for destinations that decide to impose the arrival fee.

In order to determine the appropriate tax rate (the per-passenger arrival fee) the amount of waste that each passenger generates could be taken into account. Then, using economic valuation methodologies, the economic cost of this waste could be determined. So, the final arrival fee per person per port would be determined by the size and kind of cruise ship and its employment of cleaner waste management technologies and policies. This would, in turn, require reporting by each cruise ship of its waste generation in relation to its passenger capacity.

When evaluating an incentive-based policy instrument, a series of criteria are generally used: environmental effectiveness, static efficiency, dynamic efficiency, polluter pays principle, and revenue generating ability. A regionally implemented arrival fee per capita per port of call would pass all of these criteria.

- The policy is environmentally effective because its tax base is each individual tourist on each cruise ship, and this corresponds directly with the environmental problems of marginal waste generation of cruise ships, environmental degradation from cruise ships, and congestion because of cruise tourism. If the tax rate is high enough, it could actually have the desired environmental effect of decreasing the tax base, cruise mass-tourism in the Caribbean.
- The policy is statically efficient because it is cost-effective in achieving the objective. Since ships already pay port fees, etc, to local Caribbean governments, the administrative infrastructure is already in place. Little additional administration would be needed to implement a per-passenger arrival fee for each ship. The costs would be lower than an alternative incentive-based tool such as the creation of a cap-and-trade-style market for waste permits, which would require the creation of an entire new administration system.
- The policy is dynamically efficient because it could offer an incentive to continuously reduce the environmental problem. If the arrival fee is determined based, for the most part, on the amount of waste that each passenger generates, then the ships would have an incentive to adopt new technologies and on-board policies to reduce per-passenger waste generation.
- This policy abides by the polluter pays principle because the arrival fee would be transferred directly to the tourist. The arrival fee would appear on the passenger's bill as an environmental user fee or something to that effect. The cruise line would only be the substitute, or instrument of tax collection, much like a V.A.T.
- Finally, the policy meets the criteria of ability to generate revenues to the government instead of costs. Because cruise tourism is a form of large-scale mass tourism, its revenue generating ability is significant. As World Bank observes, "even though economic rents per person may be small, the millions of visitors mean that the total economic rent associated with sites may still be substantial (World Bank, 2001).

So, a prediction of the consequences is as follows, using Belize as an example. As part of an association of governments for the environmental protection of the Caribbean, Belize would introduce a per-passenger arrival fee on each cruise ship that docks in one of its ports of call. The fee would be calculated based on the ratio of solid and liquid waste generated to the number of people on board. If it is found to be true, for example, that each passenger generates 7.7 pounds of solid waste per day on a particular kind of ship, and that this has an environmental economic cost of $5 per day as determined my economic valuation methodologies, then this $5 would be part of the arrival fee. The same would be done to calculate the cost of liquid wastes and the cost of damage done by cruise tourism to natural environments. The total arrival fee per person would be the total of these three measurements. The final fee would depend on each ship's waste efficiency, and an acceptable fee would be around the $20 that Bermuda presently charges.

Calculating the environmental cost of cruise tourism in monetary terms would be an expensive process. Although earmarking environmental tax revenues for environmental purposes creates a new set of problems and paradoxes, it may be important in this case to use part of the tax revenue for the environmental valuation process. This could avoid the implementation of an arbitrary arrival fee based on political consent, and could also hold cruise lines accountable for their waste generation by putting an economic value on their environmental impacts based on the performance of each kind of ship. This, in turn, would have the desired environmental effect of promoting newer, cleaner technologies .

CONCLUSION

In their seminal work on the economics of cruise ship tourism, Dwyer and Forsyth foresaw the possible environmental externalities. They observed, "cruise tourism will create some environmental costs, though it may also lead to environmental benefits such as improved environmental standards in shipping and waste management" (Dwyer and Forsyth, 1998). Since that time, the industry's environmental costs have been disproportionate to the environmental benefits it has induced. However, this reality can change with the implementation of carefully designed environmental policy tools. Already, there are normative-based, incentive-based, and voluntary self-regulating policies in place, but these fall short of fully addressing environmental externalities and they lack region-wide coordination. If a region-wide effort were made on behalf of all the governments of the Caribbean to implement an environmental tourism tax such as a per-passenger arrival tax, then the economic "rent" on public goods and natural resources could be more efficiently captured.

Policies such as the environmental tourism tax proposed here could have the double dividend of addressing the cause of an environmental problem by correcting its too-low price and generating much-needed public revenues for destination governments at the same time.

CRUISE SHIP POLLUTION IN THE UNITED STATES

The cruise ship industry is a significant and growing contributor to the United States economy, providing more than $32 billion in benefits annually and generating more than 330,000 U.S. jobs, but also making the environmental impacts of its activities an issue to many. Although cruise ships represent a small fraction of the entire shipping industry worldwide, public attention to their environmental impacts comes in part from the fact that cruise ships are highly visible and in part because of the industry's desire to promote a positive image.

Cruise ships carrying several thousand passengers and crew have been compared to "floating cities," and the volume of wastes that they produce is comparably large, consisting of sewage; wastewater from sinks, showers, and galleys (graywater); hazardous wastes; solid waste; oily bilge water; ballast water; and air pollution. The waste streams generated by cruise ships are governed by a number of international protocols (especially MARPOL) and U.S. domestic laws (including the Clean Water Act and the Act to Prevent Pollution from Ships), regulations, and standards, but there is no single law or rule. Some cruise ship waste streams appear to be well regulated, such as solid wastes (garbage and plastics) and bilge water. But there is overlap of some areas, and there are gaps in others. Some, such as graywater and ballast water, are not regulated (except in the Great Lakes), and concern is increasing about the impacts of these discharges on public health and the environment. In other areas, regulations apply, but critics argue that they are not stringent enough to address the problem — for example, with respect to standards for sewage discharges. Environmental advocates have raised concerns about the adequacy of existing laws for managing these wastes, and they contend that enforcement is weak.

In 2000, the U.S. Congress enacted legislation restricting cruise ship discharges in U.S. navigable waters within the state of Alaska. California, Alaska, and Maine have enacted state-specific laws concerning cruise ship pollution, and a few other states have entered into voluntary agreements with industry to address management of cruise ship discharges.

Meanwhile, the cruise industry has voluntarily undertaken initiatives to improve pollution prevention, by adopting waste management guidelines and procedures and researching new technologies. Concerns about cruise ship pollution raise issues for Congress in three broad areas: adequacy of laws and regulations, research needs, and oversight and enforcement of existing requirements. Legislation to regulate cruise ship discharges of sewage, graywater, and bilge water nationally was introduced in the 109th Congress, but there was no further congressional action.

This article describes the several types of waste streams that cruise ships may discharge and emit. It identifies the complex body of international and domestic laws that address pollution from cruise ships. It then describes federal and state legislative activity concerning cruise ships in Alaskan waters and activities in a few other states, as well as current industry initiatives to manage cruise ship pollution.

BACKGROUND

More than 46,000 commercial vessels — tankers, bulk carriers, container ships, barges, and passenger ships — travel the oceans and other waters of the world, carrying cargo and passengers for commerce, transport, and recreation. Their activities are regulated and scrutinized in a number of respects by international protocols and U.S. domestic laws, including those designed to protect against discharges of pollutants that could harm marine resources, other parts of the ambient environment, and human health. However, there are overlaps of some requirements, gaps in other areas, geographic differences in jurisdiction based on differing definitions, and questions about the adequacy of enforcement.

Public attention to the environmental impacts of the maritime industry has been especially focused on the cruise industry, in part because its ships are highly visible and in part because of the industry's desire to promote a positive image. It represents a relatively small fraction of the entire shipping industry worldwide. As of January 2008, passenger ships (which include cruise ships and ferries) composed about 12 per cent of the world shipping fleet. The cruise industry is a significant and growing contributor to the U.S. economy, providing more than $32 billion in total benefits annually and generating more than 330,000 U.S. jobs, but also making the environmental impacts of its activities an issue to many. Since 1980, the average annual growth rate in the number of cruise passengers worldwide has been 8.4 per cent, and in 2005, cruises hosted an estimated 11.5 million passengers. Cruises are especially popular in the United States. In 2005, U.S. ports handled 8.6 million cruise embarcations (75 per cent of global passengers), 6.3 per cent more than in 2004. The worldwide cruise ship fleet consists of more than 230 ships, and the majority are foreign-flagged, with Liberia and Panama being the most popular flag countries.Foreign-flag cruise vessels owned by six companies account for nearly 95 per cent of passenger ships operating in U.S. waters. Each year, the industry adds new ships to the total fleet, vessels that are bigger, more elaborate and luxurious, and that carry larger numbers of passengers and crew. Over the past two decades, the average ship size has been increasing at the rate of roughly 90 feet (27 m) every five years. The average ship entering the market from 2008 to 2011 will be more than 1,050 feet (320 m) long and will weigh more than 130,000 tons.

To the cruise ship industry, a key issue is demonstrating to the public that cruising is safe and healthy for passengers and the tourist communities that are visited by their ships. Cruise ships carrying several thousand passengers and crew have been compared to "floating cities," in part because the volume of wastes produced and requiring disposal is greater than that of many small cities on land. During a typical one-week voyage, a large cruise ship (with 3,000 passengers and crew) is estimated to generate 210,000 US gallons (790,000 L) of sewage; 1 million US gallons (3,800 m^3) of graywater (wastewater from sinks, showers, and laundries); more than 130 US gallons (490 L) of hazardous wastes; 8 tons of solid waste; and 25,000 US gallons (95 m^3) of oily bilge water. Those wastes, if not properly treated and disposed of, can pose risks to human health, welfare, and the environment. Environmental advocates have raised concerns about the adequacy of existing laws for managing these wastes, and suggest that enforcement of existing laws is weak.

A 2000 General Accounting Office (GAO) report focused attention on problems of cruise vessel compliance with environmental requirements. GAO found that between 1993 and 1998, foreign-flag cruise ships were involved in 87 confirmed illegal discharge cases in U.S. waters. A few of the cases included multiple illegal discharge incidents occurring over the six-year period. GAO reviewed three major waste streams (solids, hazardous chemicals, and oily bilge water) and concluded that 83 per cent of the cases involved discharges of oil or oil-based products, the volumes of which ranged from a few drops to hundreds of gallons. The balance of the cases involved discharges of plastic or garbage. GAO judged that 72 per cent of the illegal discharges were accidental, 15 per cent were intentional, and 13 per cent could not be determined. The 87 cruise ship cases represented 4 per cent of the 2,400 illegal discharge cases by foreign-flag ships (including tankers, cargo ships and other commercial vessels, as well as cruise ships) confirmed during the six years studied by GAO. Although cruise ships operating in U.S. waters have been involved in a relatively small number of pollution cases, GAO said, several have been widely publicized and have led to criminal prosecutions and multimillion-dollar fines.

In 2000, a coalition of 53 environmental advocacy groups petitioned the Environmental Protection Agency (EPA) to take regulatory action to address pollution by cruise ships.The petition called for an investigation of wastewater, oil, and solid waste discharges from cruise ships. In response, EPA agreed to study cruise ship discharges and waste management approaches. As part of that effort, in 2000 EPA issued a background document with preliminary information and recommendations for further assessment through data collection and public information hearings. Subsequently, in December 2007, the agency released a draft cruise ship discharge assessment report as part of its response to the petition. This report summarized findings of recent data collection activities (especially from cruise ships operating in Alaskan waters). The report was finalized and issued 29-Dec-2008 and published in the FR in 2009

CRUISE SHIP WASTE STREAMS

Cruise ships generate a number of waste streams that can result in discharges to the marine environment, including sewage, graywater, hazardous wastes, oily bilge water,ballast water, and solid waste. They also emit air pollutants to the air and water. These wastes, if not properly treated and disposed of, can be a significant source of pathogens,nutrients, and toxic substances with the potential to threaten human health and damage aquatic life. Cruise ships represent a small — although highly visible — portion of the entire international shipping industry, and the waste streams described here are not unique to cruise ships. However, particular types of wastes, such as sewage, graywater, and solid waste, may be of greater concern for cruise ships relative to other seagoing vessels, because of the large numbers of passengers and crew that cruise ships carry and the large volumes of wastes that they produce. Further, because cruise ships tend to concentrate their activities in specific coastal areas and visit the same ports repeatedly (especially Florida, California, New York, Galveston, Seattle, and the waters of Alaska), their cumulative impact on a local scale could be significant, as can impacts of individual large-volume releases (either accidental or intentional).

INTERNATIONAL LAWS AND REGULATIONS

MARPOL 73/78 is one of the most important treaties regulating pollution from ships. Six Annexes of the Convention cover the various sources of pollution from ships and provide an overarching framework for international objectives. In the U.S., the Convention is implemented through the Act to Prevent Pollution from Ships. Under the provisions of the Convention, the United States can take direct enforcement action under U.S. laws against foreign-flagged ships when pollution discharge incidents occur within U.S. jurisdiction. When incidents occur outside U.S. jurisdiction or jurisdiction cannot be determined, the United States refers cases to flag states, in accordance with MARPOL. These procedures require substantial coordination between the Coast Guard, the State Department, and other flag states, and the response rate from flag states has been poor.

Federal Laws and Regulations

Main article: Regulation of ship pollution in the United States In the United States, several federal agencies have some jurisdiction over cruise ships in U.S. waters, but no one agency is responsible for or coordinates all of the relevant government functions. The U.S. Coast Guard and EPA have principal regulatory and standard-setting responsibilities, and the Department of Justice prosecutes violations of federal laws. In addition, the Department of State represents the United States at meetings of the IMO and in international treaty negotiations and is responsible for pursuing foreign-flag violations. Other federal agencies have limited roles and responsibilities. For example, the National Oceanic and Atmospheric Administration (NOAA, Department of Commerce) works with the Coast Guard and EPA to report on the effects of marine debris. The Animal and Plant Health Inspection Service (APHIS) is responsible for ensuringquarantine inspection and disposal of food-contaminated garbage (these APHIS responsibilities are part of the Department of Homeland Security). In some cases, states and localities have responsibilities as well. For more information about any of the laws described here, see Regulation of ship pollution in the United States.

Sewage

The Federal Water Pollution Control Act, or Clean Water Act (CWA), is the principal U.S. law concerned with limiting polluting activity in the nation's streams, lakes, estuaries, and coastal waters. Under the act, pollutant discharges from point sources — a term that includes vessels — are prohibited unless a permit has been obtained. Sewage from cruise ships and other vessels is exempt from the requirement to obtain an NPDES permit.

Section 312 of the Clean Water Act seeks to address this gap by prohibiting the dumping of untreated or inadequately treated sewage from vessels into the navigable waters of the United States (defined in the act as within 3 miles (4.8 km) of shore). Cruise ships are subject to this prohibition. Under Section 312, commercial and recreational vessels with installed toilets are required to have marine sanitation devices, which are designed to prevent the discharge of untreated sewage. Beyond 3 miles (4.8 km), raw sewage can be discharged. On some cruise ships, especially many of those that travel in Alaskan waters, sewage is treated using Advanced Wastewater Treatment systems that generally provide improved screening, treatment, disinfection, and sludge processing as compared with traditional MSDs. AWTs are believed to be very effective in removing pathogens,oxygen-demanding substances, suspended solids, oil and grease, and particulate metals from sewage, but only moderately effective in removing dissolved metals and nutrients (ammonia, nitrogen and phosphorus). Section 312 has another means of addressing sewage discharges, through establishment of no-discharge zones (NDZs) for vessel sewage. A state may completely prohibit the discharge of both treated and untreated sewage from all vessels with installed toilets into some or all waters over which it has jurisdiction (up to 3 miles (4.8 km) from land).

Greywater

Under current federal law, greywater is not defined as a pollutant, nor is it generally considered to be sewage. The Clean Water Act only includes greywater in its definition of sewage for the express purpose of regulating commercial vessels in the Great Lakes, under the Section 312 MSD requirements. Thus, currently greywater can be discharged by cruise ships anywhere — except in the Great Lakes. Pursuant to a state law in Alaska, greywater must be treated prior to discharge into that state's waters (see discussion below).

Solid Waste

Cruise ship discharges of solid waste are governed by two laws. Title I of the Marine Protection, Research and Sanctuaries Act makes it illegal to transport garbage from the United States for the purpose of dumping it into ocean waters without a permit or to dump material from outside the U.S. into U.S. waters. Beyond U.S. waters, no MPRSA permit is required for a cruise ship to discharge solid waste. The routine discharge of effluent incidental to the propulsion of vessels is explicitly exempted from the definition of dumping in the MPRSA.28

The Act to Prevent Pollution from Ships prohibits the discharge of all garbage within 3 nautical miles (5.6 km) of shore, certain types of garbage within 12 nautical miles (22 km) offshore, and plastic anywhere. It applies to all vessels operating in U.S. navigable waters and the Exclusive Economic Zone (EEZ).

Hazardous Waste

The Resource Conservation and Recovery Act is the primary federal law that governs hazardous waste management. The owner or operator of a cruise ship may be a generator and/or a transporter of hazardous waste, and thus subject to RCRA rules. Issues that the cruise ship industry may face relating to RCRA include ensuring that hazardous waste is identified at the point at which it is considered generated; ensuring that parties are properly identified as generators, storers, treaters, or disposers; and determining the applicability of RCRA requirements to each. Hazardous waste generated onboard cruise ships are stored onboard until the wastes can be offloaded for recyclingor disposal in accordance with RCRA.29

A range of activities on board cruise ships generate hazardous wastes and toxic substances that would ordinarily be presumed to be subject to RCRA. Cruise ships are potentially subject to RCRA requirements to the extent that chemicals used for operations such as ship maintenance and passenger services result in the generation of hazardous wastes. However, it is not entirely clear what regulations apply to the management and disposal of these wastes. 30 RCRA rules that cover small-quantity generators (those that generate more than 100 kilograms but less than 1,000 kilograms of hazardous waste per month) are less stringent than those for large-quantity generators (generating more than 1,000 kilograms per month), and it is unclear whether cruise ships are classified as large or small generators of hazardous waste. Moreover, some cruise companies argue that they generate less than 100 kilograms per month and therefore should be classified in a third category, as "conditionally exempt small-quantity generators," a categorization that allows for less rigorous requirements for notification, recordkeeping, and the like.31

A release of hazardous substances by a cruise ship or other vessel could also theoretically trigger the Comprehensive Environmental Response,

Compensation, and Liability Act(CERCLA, or Superfund, 42 U.S.C. 9601-9675), but it does not appear to have been used in response to cruise ship releases.

In addition to RCRA, hazardous waste discharges from cruise ships are subject to Section 311 of the Clean Water Act, which prohibits the discharge of hazardous substances in harmful quantities into or upon the navigable waters of the United States, adjoining shorelines, or into or upon the waters of the contiguous zone.

Bilge Water

Bilge Water. Section 311 of the Clean Water Act, as amended by the Oil Pollution Act of 1990 (33 U.S.C. 2701-2720), applies to cruise ships and prohibits discharge of oil or hazardous substances in harmful quantities into or upon U.S. navigable waters, or into or upon the waters of the contiguous zone, or which may affect natural resources in the U.S. EEZ (extending 200 miles (320 km) offshore). Coast Guard regulations (33 CFR §151.10) prohibit discharge of oil within 12 miles (19 km) from shore, unless passed through a 15-ppm oil water separator, and unless the discharge does not cause a visible sheen. Beyond 12 miles (19 km), oil or oily mixtures can be discharged while a vessel is proceeding en route and if the oil content without dilution is less than 100 ppm. Vessels are required to maintain an Oil Record Book to record disposal of oily residues and discharges overboard or disposal of bilge water.

In addition to Section 311 requirements, the Act to Prevent Pollution from Ships (APPS) implements MARPOL Annex I concerning oil pollution. APPS applies to all U.S. flagged ships anywhere in the world and to all foreign flagged vessels operating in the navigable waters of the United States, or while at a port under U.S. jurisdiction. To implement APPS, the Coast Guard has promulgated regulations prohibiting the discharge of oil or oily mixtures into the sea within 12 nautical miles (22 km) of the nearest land, except under limited conditions. However, because most cruise lines are foreign registered and because APPS only applies to foreign ships within U.S. navigable waters, the APPS regulations have limited applicability to cruise ship operations. In addition, most cruise lines have adopted policies that restrict discharges of machinery space waste within three miles (5 km) from shore.

PRINCESS CRUISES' COMMITMENT TO THE ENVIRONMENT

Princess Cruises makes its living on the oceans, and therefore we are committed to environmental practices which set a high standard for excellence and responsibility, and which help preserve the marine environment in which we operate. Princess has defined clear environmental goals and policies, we have established strict lines of responsibility and accountability within our company for adhering to these guidelines, and we provide our shipboard staff with the equipment, expertise and training to achieve our environmental standards. Some key points about our programme are: The cruise industry is highly-regulated, and we work closely with the regulatory and enforcement organizations that govern the cruise industry's environmental practices to ensure that our policies meet the strictest environmental regulations set forth by the International Maritime Organization (IMO), and which are overseen by the U.S. Coast Guard who enforces both international and U.S. environmental laws.

In all cases, we meet environmental requirements and, in many cases, we go beyond what is required by law. And we are continuously working to identify ways we can raise the level of our environmental standards even further. Even though disposal of many types of solid waste into the ocean is permitted by law, Princess has a zero solid waste discharge policy prohibiting the disposal of unprocessed, non-biodegradable solid waste into the ocean. All solid waste is either incinerated on board ship, or landed ashore. Princess has invested millions of dollars to equip its ships with state-of-the-art environmental technology and waste management equipment in order to process the waste that is generated during a cruise.

One of the keys to a successful waste management programme is to minimize the amount of waste generated in the first place, and to recycle where possible. To minimize plastics, Princess has worked closely with suppliers to replace plastic packaging with biodegradable materials or eliminate or reduce packaging materials. In addition to having an Environmental Operations department reporting directly to the Senior Vice President of Marine Operations, each Princess ship has its own environmental officer. They are non-watch standing officers with full time responsibility for environment and occupational safety.

STATE-OF-THE-ART TECHNOLOGY

A successful environmental protection programme begins with the necessary equipment in place to efficiently manage waste. Princess has built its ships to high environmental standards, and invested millions of dollars in waste management and garbage disposal technology. This is an important area of investment for us as we consider the future of our company, our industry, and the health of the marine environment for future generations.

SHORE POWER

Princess pioneered the use of shore power in the cruise industry in 2001 when Juneau, Alaska became the first city to create a shoreside power connection. Currently nine of our ships feature this capability to turn off their diesel engines and literally "plug in" to a power supply in port. To create this power system, Princess has outfitted its ships with a custom-built electrical connection cabinet that automatically connects the ship's electrical network to the local electrical network ashore through a sophisticated system of cables, circuit breakers and control circuits. As a result, shore-based electricity runs all onboard services during the day-long calls. Shore power is currently available in Juneau, Seattle, Vancouver,San Francisco and Los Angeles. Other ports around the world are also currently examining the installation of shore power capability.

CREW EDUCATION AND TRAINING

Having environmental technology is only the first part of an effective waste management programme. The ship's crew play a key role in the program's success and Princess shipboard employees are trained not only on equipment use, but also to be environmentally sensitive. Princess conducts environmental training for crew members, who are also well aware of the strict disciplinary procedures that apply to any violations of our policy.

SOLID WASTE MANAGEMENT

Princess has taken many steps to reduce the amount of solid waste generated on board. On all our ships the non-biodegradable materials including metals, glass and plastics are recycled ashore and all hazardous waste is landed ashore. Biodegradable food waste is processed on board the ship using pulpers, grinders and dehydrators before being discharged into the ocean, at distances exceeding what is required by law. There are large refrigerated waste storerooms on our ships, which extends cruise times in inland waters without the discharge of any food waste.

SORTING AND SEPARATING PROCEDURES

A key part of our recycling and waste management programme is the proper sorting and separation of garbage that is generated on board ship. We start by separating at source in passenger areas with our recycling bins for glass, aluminum, paper products and the like, and this continues below deck where staff who are specially trained in these procedures ensure that waste items are properly recycled, incinerated, or stored for landing ashore into an approved garbage reception facility.

RECYCLING

Princess has an extensive solid waste recycling programme for our entire fleet. Recyclable materials such as aluminum, glass and tin cans are processed through the ship's equipment, packaged and stored for landing ashore and then off-loaded when the ship reaches a port that accepts the recycled materials. Not only is recycling a part of the ships' day-to-day operating procedures, but passengers are also encouraged to assist in the effort by using special recycling bins located around the ships.

PLASTIC MINIMIZATION

Princess strictly adheres to the laws which prohibit the disposal of plastic materials at sea. Over time Princess has re-designed its food and supplies purchasing and packaging requirements to cut down on the number of plastic items that are brought on board a ship. Plastic has been replaced by other biodegradable materials or eliminated altogether as much as possible. For example, laundry bags have been switched from plastic to paper that can be either recycled or safely incinerated.

ZERO SOLID WASTE DISPOSAL AT SEA

Under the environmental laws developed by the IMO in conjunction with the U.S. and other maritime nations, standards set forth in the International Convention for the Prevention of Pollution from Ships (MARPOL) allow ships to dispose of waste into the oceans depending upon the waste material and how far the ship is from shore. However, Princess operates in excess of the applicable laws, and does not dispose of unprocessed non-biodegradable solid waste into the ocean. The ash resulting from incineration is landed ashore.

GREY WATER AND BLACK WATER

Princess adheres to all U.S. and international environmental regulations regarding the discharge of both "grey water" (from showers, sinks, dishwashers) and "black water" (from toilets) into the ocean. All Princess ships are equipped with biological treatment plants that are certified by the U.S. Coast Guard as approved marine sanitation devices, and which meet the Clean Water Act standards of the U.S. The treatment facilities on board disinfect and naturally break down black water using a bio-reactor. After processing this is discharged into the sea at least 12 miles from shore, which exceeds the distance required by law. Grey water is held on board until the ship is underway and out of a port, and then is discharged into the sea in full compliance with all regulations, including EPA standards.

Eleven of our 16 ships have advanced wastewater treatment systems which use membrane filtration and ultraviolet light.

HAZARDOUS WASTE MANAGEMENT

The proper disposal of hazardous waste is of the utmost importance to Princess. Our comprehensive hazardous waste policies have been designed to carefully handle, segregate, label and off load such waste products from the ships into locally approved reception facilities. For example, batteries are collected from passengers at the photo shops. Similarly, used solvents, paints, medical waste are also labeled, stored and landed ashore. To reduce the amount of hazardous waste handled on board, we have instituted a product substitution programme which has enabled us to decrease the volume of such waste generated. We also regularly review equipment and revise operational procedures to reduce hazardous waste. Dry cleaning machines have been removed from the fleet.

AIR QUALITY

In addition to protecting the oceans, Princess' environmental commitment extends to protecting air quality as well. As we incinerate much of the combustible waste material onboard, the incinerators are equipped with sensors that continuously monitor the incinerator stack. Other measures to further reduce funnel emissions include purchasing fuel with a lower sulfur content than is required by international standards, and training our crew in operating procedures that minimize smoke emissions while maneuvering. Closed circuit TV cameras are installed on deck so that smoke emissions, if they do occur, can be instantly detected, enabling crew to provide an immediate response.

OIL POLLUTION PREVENTION

Oil spill prevention is another environmental concern for which Princess has prepared. Shipboard and shoreside emergency response plans are operational 24 hours a day. The discharge of oily water from the ship's machinery spaces is governed by strict international regulations which require such water to be either retained onboard or discharged through an oil water separator. Since all ships accumulate water in the machinery space bilge, we have taken steps to eliminate any inadvertent release of oil into the marine environment.

These include reducing the amount of water in the bilge to the lowest practical level, providing adequate holding tank capacity for the bilge water and a means to discharge it ashore, eliminating oil leaking from machinery so far as is practically possible, replacing equipment with state-of-the-art separators, and continuing to audit compliance with the regulations and make improvements as a result.

In the unlikely event that an oil leak should occur, Princess has a comprehensive emergency response plan in place to control and clean up small oil spills.

AUDITING AND INSPECTIONS

An ongoing auditing programme of our ships' operating procedures and equipment is used to identify areas where further environmental enhancements and improvements can be made. In addition to the mandatory inspections by the U.S. Coast Guard, U.S. Public Health Service, Centers for Disease Control, classification societies, and flag and port states, every Princess ship is audited by a trained staff of professional mariners who work for Princess' Maritime Compliance department.

THE ROLE OF AMERICAN NGOS IN THE REGULATION OF CRUISE SHIP POLLUTION

Each year, thousands of tons of pollution are generated on cruise ships and dumped into the world's oceans. This pollution threatens the marine environment, the cruise ship industry, and the people who depend on a healthy ocean. Neither historical nor current international oceanic regimes provide comprehensive regulation on cruise ship pollution. To address the threats posed by limited oceanic pollution regulation in American and international waters, several American non-governmental organizations have pursued diverse tactics to generate government, industry, and consumer response. To the extent that NGOsare able to influence policy, the actions of these groups challenges the realist notion that non-state actors have little influence on policy.

A case study of three prominent American NGOs demonstrates how different strategies can be implemented to influence oceanic policy. The Bluewater Network, Oceana, and Conservation International have tried to abate pollution from cruise ships by either lobbying the government for improved policies, generating public awareness of the cruise ship pollution issue, or working with the cruise ship industry to implement stronger pollution controls. These actors have responded to the weakness in the oceanic policy regime. Although cruise ship pollution remains a major threat to national and international waters, American NGOs have been successful in generating increased awareness of the issue and have been able encourage policies that will make cruises more sustainable for both the environment and the industry (Oceana).

THE DEVELOPMENT OF OCEANIC POLICY

International agreements have established the United States' water territory and their rights in those areas. A brief overview of the development of international oceanic policy is necessary to understand pollution policies for U.S. waters.

Government control of the oceans is a historically contentious issue (Galdorisi and Vienna 1997, p.13). Westphalian notions of firm state boundaries and resource rights do not easily translate to the ocean, where boundaries can contain neither pollution nor living resources (Borgese 1998, p. 109). To establish legal norms for ocean use, the principle of the "freedom of the seas" was first articulated by Huge Grotius in 1609 and has since guided modern oceanic policy. The "freedom of the seas", presented in Grotuis' "Mare Liberum" study, states that the oceans in general and the high seas in particular could not be owned by any individual or nation (Jacques and Smith 2003, p.7; Juda 1996 p. 8). In a harbinger of future international oceanic law, Grotius' principle noted that the only exception to the "freedom of the sea" applied to the three nautical miles extending from a nation's coast. This, he reasoned, should be considered the territory of each coastal nation because it correlated to the three nautical miles that a cannon could be shot (Jacques and Smth 2003, p.7). However, as the value of ocean resources became more apparent, the extent of a nation's territorial waters has become more controversial (Ibid). Today, the United States has strict jurisdiction over the area that extends 12 miles from the coast, and has primary jurisdiction over the area that extends 200 miles from the coast.

The "freedom of the seas" encompassed the legality of polluting the ocean because the ocean was believed to have an infinite capacity to absorb wastes (Van Dyke, Zaelke, and Hewison 1993, p. 3). Thus, any individual could legally pollute the ocean because, theoretically, it would not harm any other individual's use or enjoyment of the resource. Because of the seemingly inexhaustible resources of the ocean, policies to regulate activities such as fishing or pollution were viewed as a "ludicrous idea at best and an economically damaging one at worst." (Jacques and Smith 2003, p. 7). However, in the 1800s, reports of declining fishery stocks gradually began to challenge the notion of the limitless resources of the ocean (Van Dyke, Zaelke, and Hewison 1993, p. 1).

Over the last half century, the threats to the ocean have become increasingly apparent. However, the inherent complexities of regulating ocean use have hindered comprehensive and effective international policies. Agreements such as the United Nations Convention on the Law of the Sea have provided framework for strong international oceanic regimes, but have failed to reverse the threats to the world ocean (Friedheim 1993, p. 3-41). Despite the UN Convention on the Law of the Sea's distinction as "the most comprehensive international treaty negotiated before or since," the absence of key players and lack of enforceable policies has weakened its success (Galdorisi and Vienna 1997, forward). In the United States the Convention was signed in 1994 by the president but has not been adopted. A lack of involvement by major world powers such as the United States has undermined the effectiveness of oceanic regimes (Friedheim 1993, p. 3-41).

With a heightened awareness of threats to the ocean, the need for a strong national and international ocean regime has become more apparent. In the face of the United States' lack of international involvement, non-governmental organizations in the United States have pushed for stronger policies for domestic waters. The implementation of these policies would play a crucial role in strengthening the international oceanic regime (Friedheim 1993, p. 3-41). Grotius' "freedom of the seas" provides a weak foundation for modern U.S. and international policy. A shift away from the notion of the sea's infinite capacity to assimilate pollution is needed to incorporate modern understandings of the ocean's finite ability to provide resources and assimilate pollution. The pressure that NGOs have put on the government and the cruise ship industry promotes a shift towards policies and practices that consider the ocean's fragility.

Cruise Ships and Oceanic Pollution Law

Cruises are one of the fastest growing and most lucrative sectors in the travel industry. The number of passengers is expected to increase annually by eight percent for the next decade, and last year the industry reaped profits of over $1.5 billion. (Earth Island Journal). This growth has been most rapid in North America from where over 50 per cent of all cruise ship passengers depart. Each ship can carry over 5,000 passengers and crew and, in one voyage, generates hundreds of thousands of gallons of wastes, including greywater, sewage, oily bilge water, and toxic chemicals from photo processing and dry cleaning. The magnitude of this waste is comparable to that of a small city (Earth Island Journal; Oceana).

The most comprehensive United States policy to protect the oceans did not anticipate this growth in the cruise ship industry. Thus, regulations such as the Clean Water Act have numerous loopholes that allow for considerable pollution by cruise ships. For instance, the Clean Water Act regulates the discharge of pollutants from point sources, but sets forth no provisions for the discharge of sewage and graywater from vessels in domestic waters. This exception allows cruise ships to legally dump water contaminated with toxic detergent, oil, pesticides and metals into even the most sensitive of marine ecosystems. To exacerbate the situation, the cruise ship industry has been shown to disregard existing laws. In the five year period from 1993 to 1998, the EPA recorded 104 confirmed cases of illegal discharge of "oil, garbage, and hazardous wastes" (Earth Island Journal, 2000). In one case, Royal Caribbean Cruises outfitted their ships with a secret piping system that allowed wastes from photo development and dry cleaning to bypass pollution treatment equipment and enter directly into the water (Earth Island Journal, 2000). Characterized in government reports as a "fleet-wide conspiracy [to] use our nation's waterways as its dumping ground, " extreme violations such as these prompted non-governmental organizations to demand greater environmental accountability of the cruise ship industry.

The successes and shortcomings of NGOs in effecting cruise ship policies and practices are outlined in the case studies that follow. However, one piece of legislation, the Clean Cruise Ships Act of 2004, is indicative of the increased awareness of cruise ship pollution, due in large part to NGO action. The Clean Cruse Ships Act of 2004 was introduced on April 1st to Congress and the House of Representatives. The Act provides environmental provisions that apply directly to cruise ships and would help close the loopholes in existing legislation. It is sponsored by both political parties and, if passed, would promote more environmentally sensitive disposal of wastes generated on cruise ships. Included in the Act is legislation that would prohibit "discharges of sewage, graywater and oily bilge water within 12 miles of shore" and would allow the Coast Guard and EPA to collaborate on the establishment of effluent standards for sewage and graywater that is discharged beyond 12 miles of the U.S. coast. In addition, it would allow a civil action suit to be brought against anyone in violation of the Act (Bluewater Network; EPA; Oceana). This act has been strongly supported by the NGOs most closely involved in policy and public action. The future of the Clean Cruise Ships Act of 2004 will be largely determined by the ability of NGOs to build on their past efforts and generate sufficient public and government support for this new policy.

The Role of American NGOs in Effecting Cruise Ship Pollution Policy

Three American NGOs have taken different but connected actions in their efforts to influence cruise ship pollution. Their actions have focused on either government policy, consumer awareness, or cooperation with the cruise ship industry. Because these movements are still relatively new, the full impact of NGOs on cruise ship policy cannot yet be fully assessed. Their activities to date provide insights on the ability of non-governmental organizations to influence environmental policy. An examination of the different tactics that these groups have employed sheds light on the diverse means for changing oceanic policy and industry practices.

BLUEWATER NETWORK: GOVERNMENT POLICY

Bluewater Network, an environmental NGO based in San Francisco, has focused on improved government policy and oversight of the cruise ship industry. In response to growing concern over weak and poorly enforced regulations, Bluewater Network initiated a lobbying campaign to re-evaluate cruise ship pollution policy. With a focus on the environmental protection agency, Bluewater Network has pressured the government to be more responsive to the threats of cruise ship pollution. The centerpiece of their actions is a petition that was sent the EPA in 2000. This petition, signed by 54 organizations, demanded that the EPA take "regulatory action…to address pollution by cruise ships." The petition provided an overview of the loopholes in national laws and the violations by the cruise ship industry of existing laws.

This petition was successful in generating government action. The EPA responded immediately with the establishment of a commission to study cruise ship pollution policies. Directed by this commission, the EPA oversaw public hearings across the nation, developed policy recommendations, and implemented legislative tools such as suits to enforce existing regulations.

The wide support that the petition garnered by individuals and the 54 signing organizations presented a unified voice to policymakers. Successful collaboration with other interested environmental parties has given Bluewater Network authority on the cruise ship pollution issue. A public awareness campaign has been carried out in conjunction with the Bluewater Network's political involvement, but the majority of their resources continue to be devoted to working with policymakers to force greater environmental accountability on the cruise ship industry. As active players in the policy process, Bluewater Network supports the Clean Cruise Ships Act of 2004. The organization's ability to effect policy has already generated successes in the domestic ocean regime.

OCEANA: PUBLIC AWARENESS

Oceana is a U.S. based non-governmental organization that supports both domestic and international environmental campaigns. The organization is devoted to conserving the world's oceans through "policy advocacy, science, law and public education." (www.oceana.org). In response to cruise ship industry practices, Oceana initiated a campaign called "stop cruise pollution" that promotes more stringent regulation of the dumping of waste generated on cruise ships in domestic and international waters. Oceana has focused its efforts on educating the public about poor environmental control within the cruise ship industry and thereby pressuring the industry to respond to customer concerns. This campaign is primarily focused on generating a heightened public awareness of the cruise ships practices and transforming this awareness into public pressure for industry reform.

Oceana has achieved these objectives by organizing petitions and outreach. Petitions, public awareness tables, and advertisements have centered around ports in Florida where the majority of the world's cruise ships dock. The organization found that most cruise ship passengers were unaware of the pollution practices of cruise ships, and would support cruise lines that had stronger environmental policies, even if there was a slight increase in cost. One petition, signed by nearly 90,000 people, stated that customers would not travel with the cruise line carrier Royal Caribbean until the company improved their environmental record. This effort, coupled with general negative publicity, resulted in a recent commitment from Royal Caribbean to adopt advanced wastewater treatment (AWT) systems on all ships and to provide public documentation of the performance of this technology (Oceana, Royal Caribbean).

Oceana's public awareness campaigns have had success in coercing the cruise ship industry to reform their policies. In addition to public awareness campaigns, Oceana maintains programmes that lobby the government for improved cruise ship policies. Because of past successful outreach to the general public, it is likely that Oceana will be a major force in generating public support for the Clean Cruise Ship Act of 2004. Oceana's success with public campaigns has played an important role in promoting cruise ship reform.

CONSERVATION INTERNATIONAL: INDUSTRY COLLABORATION

Conservation International is a large environmental organization that draws on a variety of programmes to support their overarching goal of protecting "hotspots"— natural areas of great biodiversity. Their programmes are driven by the belief that humans can live in harmony with the natural world. In working for better cruise ship pollution practices, Conservation International has employed tactics very different from those used by Bluewater Network and Oceana. Conservation International has devoted its energies to working directly with the cruise ship industry to implement programmes for pollution control that balance environmental, economic, and social needs.

Conservation International's involvement with cruise ship policy has developed out of its Center for Environmental Leadership in Business (CELB). This programme brings experts on the environment, society, policy, economics together with industry leaders to facilitate collaboration on environmental issues. This partnership recognizes that the cruise ship industry depends on a healthy marine environment for in order to provide an enjoyable experience for their customers. Striving to provide net benefits to both the environment and industry, Conservation International's CELB is a departure from the work of Oceana and Bluewater Network.

The Center for Environmental Leadership in Business has developed a cooperative effort with the cruise industry by working with their main organization branch, the International Council of Cruise Lines (ICCL). This partnership began in late 2003 and has been successful in providing a forum for leaders in the cruise ship industry to work with environmental conservationists in producing voluntary plans for cruise ship pollution management. Prominently advertised on the ICCL's web site, the industry claims that their partnership with Conservation International supports the ICCL's "commitment to preserving and protecting the environment." This partnership has resulted in the publication of reports that analyze current practices and policies and present recommendations for future improvements. Unlike Bluewater Network and Oceana, Conservation International does not focus on public awareness or policy campaigns. The organization does not actively support the Clean Cruise Ships Act of 2004.

Because CELB's partnership is relatively new, the lasting results of this venture cannot yet be determined. non-etheless, CELB provides an important means of establishing common ground for industry and the environment. If both parties are earnest in their objectives, the partnership could facilitate more transparent operations.

CONCLUSIONS

American NGOs such as Bluewater Network, Oceana, and Conservation International have been effective in bringing attention to the environmental threats posed by cruise ship pollution. Employing various techniques, their actions have prompted government responses and industry changes. The cruise ship pollution campaign has been strengthened by the cooperation among organizations that the Bluewater Network petition facilitated. Heightened public awareness, engendered by NGO campaigns, has pressured the U.S. government and the cruise ship industry to respond. If policy changes, such as those presented in the Clean Cruise Ships Act of 2004, occur, the activities of Conservation International may allow the cruise ship industry to comply with environmental regulations while maintaining high profits. The confluence of each NGO's actions may provide the means necessary for a strong domestic ocean regime. In the absence of international cooperation, a strong American ocean policy is an important step in safeguarding the world's oceans. Various strategies have allowed NGOs to highlight cruise ship pollution as an issue that can not longer be ignored.

GLOBAL PERSPECTIVES OF CARIBBEAN TOURISM

Global forces have historically left indelible imprints on small countries. This has been especially true for island societies in the past half century. In the postwar era, their political economy has been largely defined by two worldwide forces: the juggernaut of decolonisation and the global spread of international tourism. In the first case, since 1960 over two dozen islands have achieved full political independence (Central Intelligence Agency 2002). As the terms of trade stagnated for colonial plantation staples, local elites who replaced metropolitan leadership used their new-found autonomy to create tax havens and other non-traditional activities like off-shore finance and ship registry (Baldacchino and Milne 2000). Most enduring success was achieved by capitalizing on their comparative advantage in recreational amenities. Like hand in glove, the rapid transformation of tourism into the world's largest industry-accounting for 12, 10, 9 and 8 per cent respectively of global exports, Gross Domestic Product, investment and employment (World Travel and Tourism Council 2002)-coincided with the restructuring of island economies away from traditional exports like sugar and copra towards mass tourism. These forces transformed the insular landscape across the Caribbean, Mediterranean, and Northern Pacific into the pleasure periphery of North America, Europe and Japan respectively (Turner and Ash 1976).

In the Caribbean, a host of historical and contemporary forces have shaped the gestation of tourism. In contrast to remote Pacific, Indian Ocean and African islands, the geographic proximity of the Antilles to the dominant origin markets in North America (Europe to a lesser extent) has significantly enhanced their low-cost appeal. In addition, centuries of core-periphery commerce have facilitated the islands access to foreign capital for hotel investment and to a steady supply of imported food, beverage and luxury gifts essential to satisfy affluent metropolitan vacation demand. Long-term political ties have also played a role and continue to do so since 16 of the 29 islands remain dependencies. The capital base of the tourism industry, the airports and docks and main island roadways, has largely been created by aid finance. More recently, in the late 1970s U.S. policy-makers provided tax incentives for U.S. corporations holding off-shore conventions in the region. Moreover, the 1990 revision of the Caribbean Basin Initiative (CBI) increased the duty-free gift allowance of U.S. citizens returning from the Caribbean from US$400-600 and in the U.S. territories (Puerto Rico and U.S. Virgin Islands) from US$800-1,200 (Coyle 1993). Finally, domestic economic policy has been influential. Island governments, habituated to a colonial tradition of high-volume, low-value-added monocultural exports, provided a receptive pro-market environment with generous tax incentives to establish the similar development of mass tourism. The confluence of all these forces with rising western affluence and the advent of paid vacations, package tours and jet technology combined to create the tropical island getaway as the North American vacation (or cruise) of choice.

Geography also played a part in the spread of tourism across the region. Because of proximity to the affluent U.S. Atlantic seaboard, tourism historically evolved from north to south in four waves. In the late 19th and early 20th centuries, a small stream of wealthy Americans established tourism in the Greater Antilles, primarily Cuba and Jamaica, and also Bermuda. The take-off into mass tourism took place in the 1950s and 1960s with the appearance of jet travel and the U.S. embargo of Cuba. As a result, the Bahamas, Puerto Rico and the U.S. Virgin Islands (USVI) in the north and Aruba and Barbados in the south became popular international resort destinations (Seward and Spinrad 1982). During the late 1970s and 1980s, a third wave of even more rapid growth washed across the archipelago. In the Greater Antilles, it included the Dominican Republic and the resurgence of Cuba as well as the British Virgin Islands (BVI) and the Cayman Islands. A final pulse engulfed the rest of the Lesser Antilles spreading south from the Leeward Islands to the Windward Islands, from St. Maarten and Antigua through Guadeloupe, Martinique and St. Lucia to Trinidad and Curacao. The Turks and Caicos Islands represent the most recent mass tourism destination, achieving 100,000 stayover visitors by the late 1990s.

The cyclic pace of this postwar expansion was also affected by a variety of external and internal factors. According to Bell (1993), in the 1950s and 1960s tourism did not attract wholesale government support because it was mainly controlled by external investors using black labor to service affluent whites and appeared to be neo-colonialism in modern guise. It also-attracted criticism from a vocal, newly independent intelligentsia, and from some international development agencies, because of sociocultural factors' (Holder 1993:21). Growth slowed in the early 1970s because of volatility caused by the oil crisis and socialist experiments (*e.g.*, Grenada, Guyana, and Jamaica) in the region. Since then, however, the secular decline in agriculture, continuing instability in mining, the relatively high-cost of manufacturing, and the failure of CARICOM to meet enhanced trade and investment expectations have fostered a more favourable climate for tourism promotion and growth has surged.

At the threshold of the new millennium, the economic significance of tourism has intensified in a region buffeted by the damaging contours of globalisation. These include the export of manufacturing (textiles) jobs to Mexico through NAFTA, the on-going phasing out of banana production for export through the consolidation of the European Union, and a drastic falloff in the U.S. aid since 1990. Most recently they include the aftermath of terrorist attacks in the U.S., the slowdown in global growth, and uncertainty from the threat of another war in the Middle East. In addition, dependence on tourism as the primary growth engine is occurring at a time when the non-sustainability of past mass tourism practice has become apparent. Condominiums, hotels and roadwork on steep hillsides have damaged forests and watersheds, causing erosion, silting over streams and wetlands and polluting lagoons (McElroy and de Albuquerque 1998). Mangrove forests and salt ponds have been destroyed by the placement of large-scale resorts, marinas and infrastructure along shorelines, depleting endemic species, archeological artifacts and reef systems already weakened by sand mining, yacht anchoring and sewage dumping (Wilkinson 1989). In short, just as centuries of deforestation for sugar culture resulted in the environmental devastation of island interiors, mass tourism itself has had a particularly noticeable direct impact on highly vulnerable coastal environments, such as beaches, coral reefs and mangrove forests (Weaver 1998:189).

THE CARIBBEAN IN THE WORLD

Pressure to grow Caribbean tourism is also increasing in an era of intensifying global competition. According to Harrison (2001), the past decade has witnessed a decline in traditional European and North American leaders and the rise of LDC destinations. Noteworthy are rising market shares for East Asia, the Pacific and Africa. Some of the most rapid growth is taking place in Eastern European economies in transition and especially in Southeast Asia. In the latter case, growth has been fueled by a combination of four related factors (Hitchcock et al. 1993): (1) the increasing availability of attractive packages and deep discounting, (2) the shift away from more familiar Mediterranean and Aegean destinations towards the more exotic, (3) the rising importance of newly industrialized countries (Hong Kong, Singapore, South Korea, Taiwan) as origin markets, and (4) aggressive promotion by ASEAN (Association of South East Asian Nations) countries (Indonesia, Malaysia, Philippines, Thailand).

Despite these trends, tourism in the insular Caribbean (excluding Bermuda) has out-performed the industry world-wide, and the region has increased market share. For example, between 1970 and 2000 the number of Caribbean stayover arrivals increased nearly five times from roughly 3.5 million to 17.2 million (CTO 1991, 2002; U.S. Department of Commerce 1993. Average compound growth rose 5.2 per cent per year versus 4.9 per cent for world arrivals over the same period. Annual growth was fastest during the 1980s (5.9 per cent) and slowest during the 1990s (4.4 per cent), perhaps a reflection both of tourism's maturity as well as of growing international competition. Over the three decades, the insular Caribbeans share of total stayover arrivals rose slightly from 2.23 per cent in 1970 to 2.46 per cent in 2000.

As a result of a generation of mass tourism development, the Caribbean has become the most tourist-penetrated region in the world. This is indicated by the Tourism Satellite Accounts (TSA) which aggregate direct and indirect effects of visitor spending to determine the industry's economy-wide impact. According to recent World Travel and Tourism Council Estimates (2001), tourism accounts for roughly 17 per cent of total Caribbean Gross Domestic Product (GDP) in contrast to the next most dependent regions: 12 per cent for North America, Europe and Oceania. Second, tourism accounts for over 21 per cent of all Caribbean capital formation while comparable figures for Oceania (13 per cent) and North America/Europe (10 per cent) are significantly lower. Third, regional tourism accounts for nearly 20 per cent of exports in contrast to 15 per cent for Oceania and 7-8 per cent for North America/Europe. Finally, tourism accounts for roughly 16 per cent of total Caribbean employment compared to 12 per cent or less for the other regions. In addition, theses TSA figures are likely too low since they exclude six islands in the estimates: Bonaire, Montserrat, Saba, St. Eustatius, St. Maarten and the Turks and Caicos. The World Travel and Tourism Council forecasts that the Caribbean will continue to lead the world in economic importance over the next decade (World Travel and Tourism Council 2002).

TOURISM PENETRATION INDEX

A snapshot of the comparative performance of Caribbean destinations (and the region indirectly) in the context of worldwide tourism can be developed by applying the Tourism Penetration Index (TPI) to a subset of small islands across all major oceanic basins. This technique attempts to comprehensively measure tourism development in small islands (roughly one million population or less) with the use of three standard and easily accessible impact indicators: (1) per capita visitor spending to measure economic impact; (2) average daily visitor density [(stayovers X average stay + one-day excursionists) divided by the host population X 365] per 1,000 population to measure socio-cultural impact; and hotel rooms per square kilometer to measure tourism's impact on the insular landscape and fragile ecology. The advantage of these indicators as constituted in the TPI is that they represent an integrated comprehensive measure of overall tourism development, but because of their highly aggregative nature, they remain rough approximations. As such the TPI cannot account for geographic or seasonal concentration of visitation nor capture a destination's tourism style or long-term experience with and/or adaptation to tourism

(McElroy 2002). To construct the index, standardized indices based on the three indicators were calculated by taking the value of each variable for each destination, subtracting the minimum of that value for the whole sample, and then dividing the result by the sample maximum minus the sample minimum according to the formula:

(X - Xmin)/(Xmax - Xmin)

The overall TPI scores and destination rankings were then calculated as the unweighted average of the three standardized impact indices (McElroy and de Albuquerque 1998).

The interpretation of the results is based loosely on Butler's (1980) destination (or product) lifecycle model that argues that tourism is dynamic and successful destinations normally pass through a regular sequence of growth stages that parallel the S-shaped logistic curve. More specifically, the interpretation is informed by an abbreviated three-stage version of the lifecycle comprising low-density, intermediate, and high-density levels or stages (de Albuquerque and McElroy 1992). The over-riding assumption is that Butler's evolutionary theory provides a useful framework for understanding the general processes and patterns of tourism development (Hovinen 2002).

To operationalise the index, a sample of 47 small islands, with populations of roughly one million or less, and for which relatively complete data were available, was selected. To ensure uniformity, all data were taken from standard sources: the tourism data from the Compendium of Tourism Statistics (WTO, 2002), and the population and area figures from The World Factbook (Central Intelligence Agency, 2001). The resulting sample includes 23 islands in the Caribbean, 14 in the Pacific, five in the Indian Ocean, two in the Atlantic (Bermuda and Cape Verde), two in the Mediterranean (Malta and Cyprus) and Bahrain in the Persian Gulf. Three have slightly more than one million inhabitants (Hawaii, Mauritius, and Trinidad), and all are less than 20,000 square kilometers in area except the Solomons.

RESULTS

The three impact variables, their standardized indices and their combined TPI scores and destination rankings. Although highly aggregative and only a rudimentary first approximation, the TPI loosely ranks the 47 islands from most (St. Maarten) to least (Solomons) penetrated. More important than individual destination rankings, however, are the island clusters revealed by the TPI and informed by a sense of the literature and historical observation. The sample divides roughly into three distinct levels of tourism development (least, intermediate, most) that correspond broadly to Butler's lifecycle trajectory of increasing visitation, facility scale and environmental impact. The Caribbean dominates the most developed group comprising the top six (including Bermuda) in this subsample and 70 per cent of all most developed destinations. The

Caribbean also comprises five of the top six in the intermediate group. This strong small-island international showing is noteworthy since four of the region's tourism leaders (Cuba, Dominican Republic, Jamaica and Puerto Rico), which account for over half of all visitor activity in the Caribbean, were excluded from the analysis because of their large size. On the other hand, the more remote and less developed South Pacific and Indian Ocean destinations dominate the least tourism-penetrated group.

INSERT TABLES 1 AND 2 ABOUT HERE (LANDSCAPE FORMAT)

With some exceptions, the most developed islands define a large slice of the pleasure periphery and include 13 mature destinations with an average per capita visitor spending over $10,000 and an average daily visitor density approaching 160 tourists per 1,000 residents. Visitors on average represent the rough equivalent of a 16 per cent increase in the daily population. Their landscapes are crowded with over 25 rooms per square kilometer. This group includes three clusters: (1) five traditional resort islands comprising Bermuda, British Virgins (BVI), Hawaii, Malta, and U.S. Virgins (USVI); six more recent mass tourism destinations of Aruba, Caymans, Guam, Marianas, St. Maarten, and Turks/Caicos; and (3) two small Dutch Antilles, Saba and St. Eustatius, known for their dive tourism. Although the TPI does capture the heavily built character of Bermuda, Malta and St. Maarten with over 60 rooms per square kilometer, as a land-based indicator it overstates the tourism impact in the BVI and the two tiny Dutch Antilles that cater principally to marine tourism. These latter two should more appropriately be considered intermediate destinations.

As a group, many of these most developed destinations share a relatively unique profile. According to the literature (McElroy and de Albuquerque 1992) and recent CTO data (2002), these mature, affluent areas, excluding Saba and St. Eustatius, are characterized by relatively slow population, visitor, and room growth rates. Their market is by and large dominated by shorter-staying visitors (6.6 nights on average) with strong preference for hotels, large-scale (100+ rooms) facilities, and man-made attractions. They also exhibit the highest levels of hotel occupancy, promotional spending, overseas marketing offices and employment and (for the Caribbean) cruise passenger traffic. As an indicator of their integration into the global tourist economy, they tend to display the lowest degree of seasonality through special year-round packages (*e.g.*, honeymoon weekends, conventions, carnivals, regattas). They also tend to exhibit a relatively high degree of man-made attractions (*e.g.*, casinos, golf courses, conventioneering). Many of these established destinations are among the most frequently cited in the literature for tourism-induced ecosystem damage, marine pollution, over-crowding, host tensions, and declining vacation quality (Jenner and Smith 1993; Beller et al. 1990; Towle 1985).

In addition to their climate and natural assets, these popular resort areas share geographic proximity to their major respective origin markets and long-standing and continuing commercial relationships. Historically, many functioned as colonial nodes in center-periphery trade, and these links have fostered the inflow of postwar hotel and other investment. Moreover, all except Malta are political dependencies or states (*e.g.*, Hawaii) with the benefit of enduring legal and fiscal ties. In several cases this has nourished access to aid-financed transport and communications infrastructure and other advantages conducive to tourism growth. These include for the U.S. Territories (Guam, Marianas, USVI), for example, ease of travel for American visitors (no passports, same language and currency) and special tax and duty-free gift/liquor purchase allowances not available to their politically independent island neighbors.

In contrast to the high-density group, the least developed destinations contain 13 primarily Pacific, Indian and African islands at the early stage of the resort cycle. As a group they average less than $200 in per capita visitor spending, one hotel per square kilometer, and contribute only a one per cent increment to the daily population. Many are considerably larger in population and area than the mature destinations and possess more abundant agricultural and mineral resources and more diversified economics. They cluster into three heterogeneous sub-groups. At the bottom are several low-income remote Pacific (Marshalls, Kiribati, Solomons, Vanuatu) and African (Cape Verde, Comoros) outposts with limited tourism development and characterized by heavy dependence on emigration, remittances, aid and public employment, *i.e.* the contours of so-called MIRAB societies (Bertram and Watters 1986). Towards the top are four destinations with highly developed tourist regions masked by the TPI: Fiji, Mauritius, New Caledonia and Reunion. In the middle are islands advancing along the cycle: Trinidad experiencing ten per cent stayover growth since the mid-1990s (CTO 2002) and Samoa achieving increasing international recognition as an ecotourism destination.

Many least developed islands exhibit characteristics of Butler's early tourism stages: small-scale facilities and infrastructures, limited visitor and room growth, and less penetrated ecosystems and cultures than the mature islands. They tend to spend least on promotion and have the lowest share of hotel rooms in large facilities. They also have the highest ratio of regional inter-island visitors and enjoy the longest average length of visitor stay (11 nights), *i.e.* up to two weeks in New Caledonia, Kiribati, R-union, Solomons and Tonga. Their slower progress up the resort cycle stems from a variety of factors. These include geographic isolation and limited infrastructure which have inhibited the expansion of direct air access to metropolitan origin markets. In contrast to the Caribbean, many also possess a less extensive colonial and commercial history with the west which has slowed their socio-economic modernisation and attractiveness for foreign investment. Because of their diversified economies, policy-makers have felt somewhat less pressure to aggressively promote mass tourism. Finally, in some cases (*e.g.*, Comoros, Fiji), political instability has hindered development.

The 21 intermediate destinations are the most dynamic and heterogeneous. Two thirds are Caribbean islands. As a group their average TPI scores fall cleanly between the most and least developed. For example, average per capita visitor spending is approximately $2,200, average daily density is 57 visitors per 1,000 residents, and average rooms per square kilometer is seven. Their average length of visitor stay is 8.6 nights. In many cases, these islands are characterized by very rapid visitor growth and hotel and infrastructure construction. In contrast to the most developed destinations, they tend to have higher rates of seasonality and lower levels of promotional spending. In terms of lifecycle progression, all have considerable tourism experience, display increasing facility scale and socio-environmental impact, and tend to differ in size, economic structure and tourism style. Clustered at the top are five developed Caribbean destinations plus Maldives and Cyprus. Two TPI anomalies are apparent. Bonaire is likely over-ranked because of its emphasis on dive tourism. On the other hand, Bahamas is undoubtedly under-ranked and should more accurately be considered a mature destination because of its high-density concentration of activity in the Freeport-Nassau area. However, its intermediate TPI score results from the archipelago's large land area and the lack of development in the out islands.

In the middle are a large number of diversified island economies with smaller tourism sectors in various stages of transition from dependence on traditional exports. In the Caribbean, there are St. Lucia and Grenada (bananas) plus St. Kitts/Nevis (sugar), Curacao (petroleum) and the French dependencies of Guadeloupe and Martinique, replacing sugar with banana exports to France. In the Pacific, there are French Polynesia (farming), the MIRAB economy of Cook Islands, heavily dependent on remittances and subsidies from New Zealand, and Palau, a TOURAB (see Apostolopoulos and Gayle 2002) international dive destination in transition from subsistence production and public employment. In the Indian Ocean, there are Maldives (fishing) and Seychelles (tuna).

At the bottom are two recent graduates from low-density status, Dominica and St. Vincent and the Grenadines, and Montserrat. The former have become increasingly recognized as ecotourism destinations with Dominica promoting itself as the Native Island of the Caribbean and encouraging small-scale facilities, local ownership, and an emphasis on environmental, historical and cultural attractions (Weaver 2001a: 168). The TPI ranking for Montserrat, a popular North American retirement resort, has declined since 1995 when devastating eruptions of the Soufriere Hills volcano rendered over half the island uninhabitable and spawned widespread emigration.

One of the major weaknesses of the TPI is its propensity to underestimate tourism impacts in larger islands. In the Caribbean case, for example, this excludes all of the Greater Antilles which account for the lion's share of tourism: Cuba, Dominican Republic, Jamaica and Puerto Rico. Together in 2000 these four (plus Haiti) accounted for over 50 per cent of the 16.9 million stayovers to the insular Caribbean (excluding Bermuda), over 50 per cent of the $16.4 billion in visitor spending, and nearly 60 per cent of the 207,400 rooms (WTO 2002). However, were they included in the present TPI analysis, they would be positioned in the early low-density stage of the lifecycle with per capita spending between $150 (Cuba) and $625 (Puerto Rico), daily densities between 5 (Cuba) and 15 (Jamaica) visitors per 1,000 population, and an average of one room per square kilometer. All of these destinations, however, have intensely developed tourist zones the aggregative all-island TPI masks: Veradero and Cayo Coco in Cuba, the north Puerto Plata coast in the Dominican Republic, Montego Bay and Ocho Rios in Jamaica, and the Condado area in Puerto Rico.

To remedy this TPI shortcoming, Padilla and McElroy (2003) developed regional indicators for the Dominican Republic, a good test case because it represents one of the fastest growing destinations in the Caribbean and presently accounts for over 17 per cent of all insular stayovers and tourist expenditures. The Dominican Republic also maintains the largest facility infrastructure containing 25 per cent of all insular rooms available. Provisional results indicate sharply different regional stages of tourism development. For example, Punta Cana on the east coast resembles islands at the top end of the lifecycle with average spending of $8,000 per capita and daily visitor densities of over 200 per 1,000 population. The Puerto Plata/Semana region on the north coast resembles intermediate destinations with per capita spending of $2,300 and 60 daily visitors while the two southeast regions (Santo Domingo and La Romana/San Pedro) are similar to least penetrated islands. Analogous results are expected for the other three dominant destinations when regional data becomes available.

NEW DIRECTIONS

To maintain its position in the global tourist economy, the Caribbean will have to devise alternative strategies to traditional mass tourism promotion that emphasize quality over quantity. This is particularly important in a region characterized by relatively high travel costs and room rates, high labor and utility costs, and low profits and income multipliers (Tewarie 2002). As a first approximation, the TPI can function as an early warning signal indirectly identifying threats to sustainability along the three major stages of the resort cycle. For mature destinations, the key challenge is to sustain vacation quality. This will require restoring damage and preventing encroachment in fragile areas, better managing visitation through space and time, and expanding average visitor stay and quality by designing nature, culture and heritage attractions. Bermuda's experience is instructive. Stagnation and complaints of saturation in the 1980s prompted a long-term policy reassessment that resulted in controls on bed capacity, cruise ships and construction design. Since then tourism has stabilized and natural and heritage amenities have been protected (McElroy 2001).

For highly developed intermediate destinations, the key issue is controlling growth within insular socio-economic and ecosystem absorptive capacities. This will require containing further renewable resource losses, sequencing large developments over long time horizons, and providing residents not only with decision-making participation but also with a viable financial stake in the industry. This last can be achieved by using incentives to foster small-scale, local, labor-intensive businesses, local purchases by hotels/restaurants and developers, and targeting regional tourists whose spending per dollar has greater domestic impact (Sinclair and Voker 1993). A ten per cent increase in the tourism income multiplier maintains economic contribution, with roughly ten per cent fewer visitors and socio-environmental impact. Here the experience of Antigua is useful. Between 1975 and 1990, total visitors doubled twice and more coastal ecosystems and habitats were damaged than in all previous history (de Albuquerque and McElroy 1995). Since 1990 stayover visitors have been volatile and/or flat. In a reverse of the Bermuda case, Antigua's failure has been largely due to a tradition of environmental neglect, marginal citizen participation, and a persistent policy preference favouring short-term economic gains over long-term environmental stability.

For less developed intermediate and low-density destinations, the key challenge is achieving commercial viability. Ideally this demands a long-term participatory planning process to identify unique assets, to construct infrastructure necessary for sustainability accessing these assets, and to devise a compatible destination identity to promote.

This is a time-consuming task but such islands have room to maneuver because of their early position in the resort cycle and their less exploited natural and cultural amenities. They also have the greatest opportunity for developing ecotourism attractions. Some of the more successful examples include the Hol Chan Marine Reserve and Chao Creek Terrestrial Reserve on the Belizean mainland, the reknown marine parks in Bonaire and Saba, the expansion of heritage sites in St. Lucia and Cayman Brac, the USVI National Park in St. John. Clearly the best example of a deliberate ecotourism destination is Dominica, promoting a combination of its riverine ecology, volcanic assets and Carib culture (see Weaver, this volume). However, managers must remain constantly vigilant lest success breed pressure for unsustainable visitation at such delicate sites. According to France and Wheeler (1995:68), they should not be seduced by the myth of green tourism, which may be little more than a growth policy masquerading behind the psuedo respectability of a green mantle. The critical factor in the 'success' of any ecotourism development is effective control of numbers.

OTHER CHALLENGES

There are a number of other external and internal challenges the Caribbean must confront to retain its premier position in global tourism. A brief sampling of the more serious deserves mention. First, in the wake of September 11 and the employment losses and eroded profits from deep discounting, the region must ensure adequate transport security infrastructure is in place to avoid further pull-outs like the discontinuation of two British cruise ships to Trinidad and Tobago (Jamaica Gleaner 2003). Second and related, 2002 was one of the worst years in airline history as evidenced by bankruptcy threats from mainland carriers and substantial losses suffered by major regional carriers like Air Jamaica, BWIA and LIAT. Since air transport is the bone marrow of the long-haul tourism typifying the Caribbean, concerted efforts must be made to explore long-term partnerships with both mainland and island airlines to consolidate service and stabilize visitor access to the region.

Third, although their overall economic impact is considerably lower than stayovers, cruise arrivals represent the fastest growing segment of the industry and will soon rival the hotel sector in bed/berth capacity. Because of the security, comfort, high satisfaction and vacation variety they offer, cruise ships have become very competitive with land-based tourism. To maximize their contribution, better partnerships must be established between cruise lines and island governments to reduce the impact of their sewage and other debris on the delicate marine environment (Ocean Conservancy 2002). In addition, their regional impact can be enhanced with greater on-land purchases from island suppliers and more cooperative 'fly-cruise' packages that promote overnight hotel stays as is presently underway in Bermuda, Jamaica and elsewhere. Fourth, for their part, regional decision-makers not only need to appreciate the increasing significance of tourism is their policy-making, but also to recognize the long-term value of cooperatively developing uniform region wide promotion, incentive and regulatory regimes to reduce destination competition with foreign (*e.g.*, air, cruise, hotel) suppliers. The recent mounting of the first major Caribbean marketing campaign is a promising step (Harewood 2002).

Finally, given the importance of visitor safety, crime is perhaps the most systemic internal threat to Caribbean tourism. Periodic reports of cruise ship cancellations because of robberies against passengers and crew underline the problem (Schladen 2002). Although the nexus between crime and tourism is becoming clearer-lucrative targets lacking precaution in unfamiliar surroundings less likely to report and/or return for trial (de Albuquerque and McElroy 1999a)-the relationship is complicated by the increasing significance of the region as a primary narco-traffic route for cocaine transshipment from Colombia to the U.S. (Drug Enforcement Agency 2001). Although violent crime is highest in many of the most mature destinations (de Albuquerque and McElroy 1999b), much is linked to turf wars among rival posses (gangs) as well as the deportation of Caribbean criminals from the metropolis: 'they leave our islands as high school criminals and return to us as post graduates' (Commonwealth Secretariat 1997: 105). Whatever the source of rising criminality, in the intermediate run Caribbean leaders must redouble efforts to strengthen enforcement with both local and extra-regional resources, and in the longer run address the more daunting task of creating a more egalitarian tourism in which all social strata share a viable economic stake and the motivation to sustain it.

SUSTAINABLE TOURISM AND CRUISES

The cruise industry is one of the fastest growing economic sectors with constant high annual growth rates worldwide. In 2000, tourism generated - direct and indirect – about 12 per cent of global GDP and nearly 200 million jobs. Earnings from international tourism in 1997 were 443 billion US dollars and are expected to reach 2 trillion US dollars in 2020. The number of people travelling internationally increased from about 463 million to 594 million per year from 1992 to 1997 (1).

International tourism generates different effects – the impacts are positive, but also negative. One of the most crucial negative effects are the impacts of tourism on the environment and the perspectives for more sustainable developments in tourism destinations. Positive impacts include positive effects on employment and income in tourism destinations, but also on the tourism industry as supplier of tourism related products and related businesses worldwide.

Tourism destinations as well as tourism industry both depend on natural environment quality – but also on a positive economical and social environment surrounding the tourism centers. But awareness is rising regarding the fact that the positive impacts of tourism can only be fostered for the future, if the negative effects of tourism on ecological, economical and social environments can be limited significantly (or avoided totally) today. The guiding principle of a more "sustainable tourism" is gaining importance.

GUIDELINE "SUSTAINABLE DEVELOPMENT"

Reference system for sustainable tourism is the principle of sustainable development.

Accoring to a definition put forward by the World Bank, sustainable development can be characterized by three principles:

- The rate at which renewable resources are used must not exceed the rate at which they are regenerated;

- The rate at which non-renewable resources are used must not exceed the rate at which renewable substitutes are developed;
- The rate of emissions of polluting agents must be in accordance with the environmental capacity to assimilate them.

A more general guideline, which includes the criteria above, but reaches beyond implies: All social, economical and ecological development approaches of the people living today should not restrict the development perspectives of future generations. Transferred to tourism issues, this principle of sustainable development implies three components (2):

Ecological Sustainability

This is about the contribution of tourism to more preservation and protection of ecosystems worldwide. Intact nature habitats are and will be an important precondition and the basis for tourism activities. Aware of this, strategies need to be developed on how to conserve resources and to reduce emissions. It also needs an environmental planning concept, which guides the way to a more environmental sustainable tourism and related activities on all levels.

Social and Cultural Sustainability

This is about the contribution of tourism to more intercultural exchange and mutual understanding among people, despite different nationalities, languages and religions. In the destinations, tourism should enable self-determined cultural dynamics and contribute to social contentment. Tourist are guests in dedicated regions and the regions are not museums. First of all, regions are the habitats for the local inhabitants, and not service providers for the tourism industry. Social responsibility implies that tourism should support the conservation of social and cultural values locally.

Economical Sustainability

Sustainable tourism is guided by the principle of a qualitative economical growth. In the short and long run, this principle contributes to diversified incomes - taking into account the ecological and social costs and benefits of growth. Tourism can contribute to the welfare of destination regions as protecting local sources of incomes and improvements of local living conditions. Additionally, tourism can contribute to facilitating disadvantages or even to equalling structural disparities of remote areas. Tourism needs to be integrated in the local economical structures in order to maximise its contribution to the regional value added. The implementation of the most comprehensive social and legal standards of most developed states has to be accepted and supported in the tourism destinations. Social and environmental dumping should not be allowed to pay off.

What has been formulated above for tourism in general, must also be true for all sectors of tourism – including cruise tourism. Activities connected to cruise tourism affect economic development and employment worldwide. But they also have impacts to the oceans and seas as well as to remote islands where they affect local development perspectives (3).

ECONOMIC ASPECTS OF CRUISE TOURISM

Cruise tourism is an important economical sector of tourism. The total value added of the US cruise industry is estimated to 12 billion US dollars, with expected growth rates of 8 per cent per year. In 1998, about 223 cruise ships carried almost 10 million tourist. From 1993 to 1998 the number of ships, operating from US-American ports increased by 50 per cent. In early 2001, 64 new – even bigger – ships were on order – a significant share of this new vessels were ordered from European ship yards. Until 2004, cruise capacities worldwide are expected to grow by 35 per cent. More than 50 per cent of all cruise ships are dedicated to the US-American tourism market and are operating in the Caribbean, on average, these vessels carry about 2000 passengers. The biggest units operating in the Caribbean today have a carrying capacity of 5000 people (including crew). Some ports of call have 15000 tourist every day which implies significant – not always positive - impacts on the social and economical structures in the destination regions.

CRUISE TOURISM AND CONSUMPTION OF RESOURCES

Cruise tourism needs resources and generates liquid and solid waste. On a one-week trip, an average Caribbean cruise ship generates about 50 tons solid waste, 7,5 mio liquid waste, 800.000 litres sewage and 130.000 grey water. About 75 per cent of the liquid wastes of ships are caused by cruise ships. From 1993 to 1998, about 100 environmental incidents caused by cruise ships were officially registered. The true figure of incidents is expected to be much higher, since liquid waste disposal happens off shore outside controlled national coast areas or within coast areas of states, which accept waste disposal for a fee (4).

Additionally, cruise tourism requires certain infra- and suprastructure in the destination regions. The investments needed to meet the needs of the cruise industry often exceed the financial capability of the destination regions. Once having invested in adequate facilities, they often depend on the cruise industry for decades.

SUSTAINABLE CRUISE TOURISM

Cruise tourism covers various segments of different markets. In most parts of the cruise business, sustainability aspects – in terms of ecological sustainability – are already emphasised. It can be expected that the cruise industry will continuously work on diminishing all negative ecological effects of cruise tourism – in there own interest.

Therefore, the supposition that, *e.g.* activities related to the most important Caribbean cruise market do not sufficiently fulfil sustainability criteria, can be shown in the social cultural as well as economical dimensions of sustainability. But it is not easy to find clear evidence (5). There is no doubt: cruise tourism stimulates local economies. But very often, the real impacts to the economical development in the long run lag well behind the promises and expectations. The relation between input and output, between costs and benefits seem unbalanced - to the disadvantage of the cruise destination areas. Unfortunately, a clear quantification of this imbalance is not available.

6

Geographical Overview of the World Cruise Market and its Seasonal Complementarities

THE CRUISE MARKET: NOT JUST FOR THE MIDDLE-AGED AND ELDERLY

With growing interest from Asia, the cruise industry is speeding ahead. Sally White reports on this fast-growing sector In an industry with a $7.6 trillion global valuation, a figure of $117 billion might attract the description of 'mere'. Cruising is as yet a tiny fraction of the World Travel and Tourist Council's figure for the vast global travel and leisure industry, but it should by no means be dismissed.

Investors certainly haven't overlooked it! Shares in industry major Carnival have soared from $33 to $55 in New York over the past 12 months and from 2,093p to 3,576p in London. (Carnival is planning on growing its 10 million passengers to an estimated 13 million by 2022.) No. 2 operator Royal Caribbean Cruises has seen its shares go from $52 to $93 in New York.

Already strong growth will be given extra impetus by the 22 new ocean, river and specialty cruise ships hitting the water this year, the 17 more scheduled over the next two years, the ten new cruise ports planned in China by 2020 and a host of other major investments. China has even ordered its state companies to build cruise ships. The industry's own major body, the Cruise Lines International Association (CLIA), says that globally preliminary ship orders extending to 2022 are valued at $26 billion.

CHINA HAS EVEN ORDERED ITS STATE COMPANIES TO BUILD CRUISE SHIPS

Only the cruise industry's ability to woo online customers is questioned by analysts who list this as a possible hobble on growth. Cruise Market Watch figures for the cruise industry give it 'an annual passenger compound annual growth rate of 6.55 per cent from 1990 – 2019'. Yet the market is very far from saturated. Only 23 million passengers are expected to cruise this year. Most holidaymakers have yet to go on a cruise – Cruise Market Watch says just 24 per cent of North Americans, for example, have ever taken a cruise. To give some perspective, it offers this – 'All the cruise ships in the entire world filled at capacity all year long would still amount to less than half of the total number of those going to Las Vegas'.

The online qualms come from perceptions that cruising is for the middle-aged and elderly, and online booking the domain of the young. The CLIA, seems to acknowledge this as it says: - 'Travel Agents Are Key to Cruise Travel – While the internet and mobile devices have overtaken how consumers make purchases, travel agents continue to be the most popular and best way to book a cruise. In fact, seven out of ten cruise travellers (70 per cent) use a travel agent to plan and book cruise vacations.'

However, according to a CLIA survey carried out just last year for its Cruise Market Profile: 'The average age of cruisers was 49, with age demographics spread fairly evenly across the spectrum.' Of course that age is not exactly 'young' but neither is it geriatric and it usually comes with a higher salary. Another important factor for the industry is that maintenance of the necessary high standards does not come cheaply; the CLIA puts the average cruise price at $1779.82 per person.

Digital marketing agency Inside Online took a look at search volume on 20 of the top companies' web sites over the last year and in its view is that five were 'underperforming'. After monitoring the sites Inside Online came up with the conclusion that the web site with the largest search growth was www.royalcaribbean.co.uk (the No 2 market leader) with an increase of 171 per cent in organic search in the 12 months to September 2015. The hardest hit was www.iglucruise.com with a decrease of 43 per cent.

(The top three cruise companies are Carnival Corporation, Royal Caribbean Cruises and Norwegian Cruise Line Holdings. They account for 81.6 per cent of worldwide share of passengers carried and 76.7 per cent of worldwide share of revenues, according to the CLIA.)

FAMILY FRIENDLY PUSH

The cruise companies are working hard to improve their online marketing, putting on their sites virtual reality views of the luxuries and services on their boats and the scenic destinations. The shots include a lot of young families! They are also installing the fastest wifi and encouraging use of social media. An Amadeus survey found that technology is very important to the cruiser with 62 per cent of respondents expecting 'super-fast' wifi onboard the ship and 30 per cent requiring 24/7 technology support throughout the journey.

Cruise companies wanting to grow should be pushing on an open door! Interest in cruise holidays seems to be booming all round the world, in Asia especially.

Market shares, according to Cruise Market Watch, are expected this year to be 58.6 per cent North America (Canada, United States and Mexico) followed by Europe (25.9 per cent), Asia (8.5 per cent) and Australia (4.3 per cent). But by 2019, 25.3 million cruise passengers are expected to be carried worldwide of which 55.8 per cent will originate from North America, 25.1 per cent Europe and 19.1 per cent the rest of the world.

CHINA AND INDIA ARE NOT ONLY GROWTH DESTINATIONS, THEY ARE ALSO SEEING DYNAMIC DEMAND.

Enviable growth is being seen in CLIA's specialty segments - sophisticated ships, luxury yachts, elegant ocean liners and the newest river cruises. It says specialty cruises grew by 21 per cent annually in 2009-2014. Top destination has long been the Caribbean which has won a third of global deployment capacity market share in 2015. However, the Mediterranean continues to grow as a destination, as well as Asia and Australia. This year 52 ships are providing 1,065 Asian cruises with capacity for 2.17 million passengers.

China and India are not only growth destinations, they are also seeing dynamic demand. "As tourism in Asia grows, particularly in affluent travel in the region, opportunities are rife across the various sectors of the tourism business," said Amrita Banta, managing director at Singapore-based international research group Agility Research and Strategy. "One of these growth areas is the newly evolving cruise industry." Research in the luxury market this year found that 68 per cent of Chinese and 61 per cent of Indians were considering taking a cruise holiday.

She added: "Indian cruises can fit into more compact packages.... but cruises geared towards the Chinese market will require more days. While cruises from Bali – Sydney are already available, these typically run up to around nine days." The Chinese interest is excellent news for both the cruise companies and the OTAs. China is not only the fastest growing travel market, (according to international consultants AT Kearney), its consumers also prefer to do business online!

CRUISE MARKET OVERVIEW

According to CLIA Cruise Line Member Survey statistics, the list of top passenger source countries includes: USA (51,7 per cent), UK (8,1 per cent), Germany (7,7 per cent), Italy (4 per cent), Australia-NZ (3,6 per cent), Brazil (3,4 per cent), Canada (3,4 per cent), Spain (2,8), France (2,4 per cent) and Baltic countries (1,6 per cent).

The top cruise destinations list shows as major market opportunities the regions of Caribbean, Asia, Europe (Mediterranean, Holy Land-Middle East), Australia, European rivers, US rivers, South America, Antarctic, World Cruises, Canada-New England, Africa. For the cruising market related professional studies and reports, you can visit the CLIA's web site cruising.org.

MAJOR CRUISE COMPANIES

Based on all existing cruise ship orders and corporate brand-line transfers, by 2022, the largest cruise line companies in the world (Carnival Corp, Royal Caribbean, MSC Cruises, NCL Norwegian) will account for 216+ vessels in total. These four major cruise companies should have a combined passenger capacity of 25,7+ million. Their estimated total capacity for 2015 is 18,7 mill passengers, with combined fleet of 180 vessels.

Carnival Corporation's revenue	• 2014 – USD 15,46 bn • Carnival Corporation brand lines are: AIDA, Carnival, Costa, Cunard, Holland America, P&O (UK and AU brands), Princess, Seabourn.
Royal Caribbean revenue	• Royal Caribbean Cruises Ltd (RCCL) worldwide revenue in 2014 was USD 8,07 bn (compared to the USD 7,96 bn in 2013). • RCCL brand lines are Royal Caribbean International, Celebrity, Pullmantur, Azamara and CDF. • RCCL holds 50% stake in TUI Cruises. • Expected number of passengers carried worldwide (by all RCCL brand lines) for 2015 (~7 mill) and expected growth for 2022 (~8 mill).
Carnival Cruise Lines worldwide passenger capacity	• *Princess Cruises (Carnival Corporation's brand line) had an ~41,000 daily passenger capacity in 2014. • Expected number of passengers carried worldwide (by all Carnival brand lines) for 2015 is ~10 mill. The expected growth for 2022 is ~13 mill.
NCL Norwegian Cruise Lines worldwide passenger capacity growth	• The NCL's percentage growth in carried passengers is the most impressive. Between 2015 and 2022, the increase in capacity will be by ~50%.
MSC Cruises	• The privately owned MSC cruise company has grown in passenger capacity by the impressive 800% since 2004. In 2014, this line carried 1,67 mill passengers, with generated profit of ~1,5 billion Euro. The forecast for 2015 shows an expected growth of 10%.

CRUISE LINE MARKET SHARE

The 2015 cruising industry worldwide is estimated at USD 39,6 bill (which is an increase of 6,9 per cent over 2014) with 22,2 mill passengers carried (which is an increase of 3,2 per cent over 2014). The cruise line market share (all brand lines) is as follows:

NOTE: In brackets are shown per cent total passengers / per cent revenue increase in US$. Carnival Corporation brand lines

(total passengers 48,1 per cent; total revenue increase 42,2 per cent)

Carnival	21,3	8
Costa Crociere	7,4	6,7
Princess	7,9	8,8
AIDA Kreuzfahrten	3,7	3,4
Holland America	3	4,4
P&O Cruises UK	1,7	3,7
P&O Cruises Australia	1,2	2,3
Iberocruceros	0,8	1,6
Cunard	0,9	2,6
Seabourn	0,2	0,7

Royal Caribbean brand lines
(total passengers 23,1 per cent; total revenue increase 22,1 per cent)

RCI	16,7	14,2
Pullmantur	1,6	1,2
Azamara	0,2	0,7
Celebrity	4,2	5,7
CDF	0,5	0,4

NCL CORPORATION BRAND LINES

(total passengers 10,4 per cent; total revenue increase 12,4 per cent)

NCL	9,5	8,7
Oceania	0,6	2,3
RSSC Regent	0,3	1,5

OTHER CRUISE LINES MARKET SHARE

(in brackets per cent total passengers / per cent revenue increase in US$)

MSC Crociere	5,2	4,2
Disney	2,8	2,4
Thomson Holidays	1,3	1,8
Hurtigruten	1,4	1,5
Louis / Celestyal	0,9	1,2
Phoenix Reisen	0,5	0,7
Fred Olsen	0	0,1
Saga Holidays	0,4	0,5
Silversea	0,4	1,3
Cruise Maritime	0,5	0,5
Crystal	0,3	1,5
Hapag-Lloyd	0,3	0,4
Ponant	0,1	0,6
Star Clippers	0,1	0,1
Lindblad	0,1	0,2
SeaDream	0	0,2
Star Cruises (Asia)	1,3	1,8
TUI	1,3	1,8
Viking (ocean + river)	18,4	23,3

CRUISE TRAVEL MARKET CHALLENGES

- Attracting first time travelers is the "deal of all deals" for the world's largest cruise companies. Because they simply cannot fill their old and new ships by stealing customers from rival lines. Statistics on this point out, that if a first-time cruiser books a particular line, she/he will buy again its product in the future. Many first timers even book their next vacation while still on the ship.
- The berths number is hugely increasing with many and big-size passenger vessels being built each year. Overcapacityleads to hugely discounted cruise deals as lines try to fill up their ships. Cruise prices go down or tickets become "more inclusive" than ever. Nowadays, cruise line industry invents and introduces new "price inclusive" onboard attractions, and to keep the revenue, lines increase prices on drinks and specialty dining on board.
- Air passenger duty has a major impact even on the fly-cruise deals market (discounted flight and cruise packages), especially UK to Caribbean, and USA to Europe (Mediterranean or Baltic). According to CLIA UK data, in 2012, the decrease of UK citizens booking Caribbean fly-cruise deals was 21 per cent.
- Still too many vacationers think of cruising vacations as expensive or boring. To battle these negative perceptions, lines generate many and "new age" adds, sponsor movies and TV series, invite pop icons to become godparents of the newest largest ships.
- According to CLIA Travel Agent Survey statistics, best motivators for booking a cruise are: Value/Price (86,6 per cent), Itineraries (77,7 per cent), Cruise Line Reputation (76 per cent), Departure Port (61,3 per cent), Amenities (39,3 per cent).

AUSTRALIAN CRUISE MARKET

In May 2015, the CLIA official report showed the Australian cruising market reaching in 2014 over 1 million passengers (1,003,256 pax) – the world's fastest growing market in one year period. This is the first destination/nation exceeding 4 per cent market penetration in the industry's history. In numbers, this is an increase of 20,4 per cent Australian passenger (around 170,000 more compared to 2013 statistics. Fun fact is, that the pre-set goal was to hit the 1 mill pax by 2020. Now the new goal for 2020 is 2 mill AU passengers, which will be a staggering 7,6 per cent market share increase. This data suggests an average annual growth rate of 12,5 per cent.

The CLIA report also showed that ~4,2 per cent of the Australia's population took an ocean ship cruise in 2014. In comparison, the North American market (ranked second) had market penetration of 3,4 per cent. The Australian cruise market is currently the 4th largest globally, with 4,5 per cent of all the world's cruise passengers, and following only North America (54,2 per cent pax), Germany (8 per cent pax) and UK (7,4 per cent).

CARIBBEAN CRUISE MARKET

Will cruise ship lines change allegiance from Caribbean to Asian regions? According to cruise trends 2015 data, there will be over 20 million customers booked on Caribbean Sea voyages in 2015.

In 2014, the industry's capacity in the Caribbean region was increased by 13 per cent, with several new big-capacity ships home-ported in the US, and mostly in Florida ports. The Caribbean will always be the most important cruise travel destination for the market of North America. Even the Italian/ Mediterranean line MSC Cruises sneds its newest and largest ship MSC Seaside for year-round operations there, with homeport Miami FL. Bottom line is – the Caribbean cruising is here to stay.

Cruise ship companies don't create demand'- they simply go where their customers want to go, where the best money opportunities are. This is the reason why changing ship deployment became a routine tactic these days. 20 years ago, it was not typical at all for cruise shipping companies to move around vessels so much. Back then, itineraries were planned like 2 years in advance, and no one ever thought about changing them.

EUROPEAN CRUISE MARKET

According to official 2015-2016 statistics, the MSC cruise line company dominates the Europe's cruising market with an estimated capacity of nearly 1,1 mill passengers. The second largest cruise line in Europe is Costa Cruises (estimated capacity of 912,500 passengers). Next in line are the German line AIDA Cruises (estimated 612,000 pax) Royal Caribbean (estimated 422,400 pax), NCL Norwegian (estimated 314,200 pax), P and O UK (estimated 276,000 pax), the German line TUI Cruises (estimated 241,500 pax). The last among the "biggest players" in Europe is Celebrity Cruises with estimated 241,500 passengers.

MSC Crociere line's most capacity is in the region of Mediterranean Sea. The AIDA Kreuzfahrten line is at leading position in North Europe and the Canaries. Compared to previous years, Royal Caribbean International reduced its fleet in the Mediterranean from 526,000 pax (2012) to 301,000 pax (2015). RCI maintains ~90,000 pax marker share in North Europe. Costa Crociere starts moving more and more ships on the Asian market. This tendency will possibly increase the MSC dominance in Europe.

CRUISES TO CUBA

Cuba as cruise destination is now open for USA travelers. Cuba promises something new, something special, something more pristine, something previously off limits. Fact is, that many people do not travel by cruise ship until they have a good reason besides the voyage itself. So along with the specially themed cruises and group travel deals, Cuba is one of the good reasons to book a ship travel vacation.

Cruise Market Statistics

Cruise industry revenue worldwide	• 2008 – 27,56 bn • 2009 – 24,93 bn • 2010 – 26,85 bn • 2011 – 29,4 bn • 2012 – 34,54 bn • 2013 – 36,27 bn • 2014 – 37,1 bn • 2015 – 39,6 bn
Number of passengers carried in Europe	• 2014 – 6,57 mill
Asia cruise passengers	• 4 out of 10 are aged 40-. • 9 out of 10 cruise within Asia. • They prefer short cruises: in 2014, 48% of them booked 4 to 6-night sailings, 38% booked 2-3-night cruises, 12% booked 7 to 13 night sailings. • Most of them are multi-generation families. • They like to do onboard high-end shopping. • They like the regional cuisine.
Avrg revenue PP worldwide	• The number of cruising people worldwide has increased from 15,62 mill (2007) to 21,12 mill (2013). The expected number for 2018 is 24+ mill. • In 2014, every cruise passenger brought ~USD 2,123 in revenue to the global industry. • The largest proportion of passengers is for the North American market. • The largest is the North American market's share of the global industry USD 21+ bill. • The statistic shows an avrg of USD 184 profit made per cruise passenger worldwide in 2014.
World's largest cruise ships weight (GT-gross tonnage)	• Allure/Oasis of the Seas (2010 / 2009) – 225,280 • Quantum of the Seas (2014) – 167,800 • MSC Vista project (Meraviglia 2017 / 2019 / 2020) – 167,600 • Norwegian TBA1 / TBA 2 (2018 / 2019) – 164,000 • Norwegian Escape/Bliss (2015 / 2017) – 163,000 • Norwegian Epic (2009) – 155,870 • Freedom/Liberty/Independence of the Seas (2006 / 2007 / 2008) – 154,400 • MSC Seaside project (Seaside 2017 / 2018 / 2021) – 154,000 • RMS Queen Mary 2 (2004) – 148,530 • Norwegian Breakaway/Getaway (2013 / 2014) – 146,600 • Royal/Regal Princess (2013 / 2014) – 142,230 • MSC Divina/Preziosa (2012 / 2013) – 139,400

North American Cruise Market

- In 2013, the North American cruising industry contributed USD 20+ billion directly to the US economy.
- ~148,000 people are currently employed by North American lines, and receive USD 6,6+ bn in wages.
- The market is set to expand with 20+ new ships with operations in North America planned for the 2014-2017 period.
- The 2007-2013 increase in available bed days is from 84,8 mill to 123,2 mill.

- 2013 statistic shows 46,7 per cent of the travel agents stated their luxury cruise deals bookings had increased over the 2012.
- A 2015 survey shows that 57,6 per cent of the respondents had experienced cruise disruption due to itinerary changes, and 6,7 per cent had experienced disruption due to quarantine.

Expenditure of the industry in the USA	• 2006 – 16,37 bn • 2007 – 17,37 bn • 2008 – 17,8 bn • 2009 – 15,98 bn • 2010 – 16,82 bn • 2011 – 17,59 bn • 2012 – 18,29 bn • 2013 – 18,3 bn • 2014 – 18,72 bn
Market volume of the industry in North America (revenue in USD)	• 2013 – 17,6 bn • 2014 – 21,2 bn
Number of passengers carried	• 2014 – 14,2 mill
Number of embarkations from USA ports	• 2007 – 9,18 mill • 2008 – 8,96 mill • 2009 – 8,9 mill • 2010 – 9,69 mill • 2011 – 9,84 mill • 2012 – 10,1 mill • 2013 – 9,96 mill • 2014 – 10,09 mill
Number of cruise ships operating the market in North America	• 2007 – 159 • 2008 – 161 • 2009 – 167 • 2010 – 169 • 2011 – 173 • 2012 – 177 • 2013 – 178 • 2014 – 185

ASIAN CRUISE MARKET

Asia's cruise industry is booming, and with China dominating the local market there is no surprise the world's major cruise lines decide to reposition ships for year-round operations there. According to the latest reports, China makes up almost 1/2 of the entire Asian cruising market.

Typically, Asians take shorter vacations in comparison to North Americans. Other statistics show that nearly 1/2 of all Asian cruise passengers are younger than 40. Asian cruise shipping lines are currently adapting by primarily offering sailings shorter than a week. From 2012 to 2014, the Asian market has nearly doubled, with plenty of room for further growth. 2015 statistics show that 1/12 of the China's population traveled in 2014, with less than 1 per cent of those traveled by cruise ship.

From March 2015, Chinese cruise passengers travelling to Japan enjoy visa-free entry. The list of operators benefiting from the new agreement includes

Carnival Corp, RCCL/Royal Caribbean, Oceania Cruises, SkySea Cruises, Bohai Ferry, HNA. Passengers leaving from China can now do cruise bookings only 3 days out. In comparison, Japan visa approval requires ~7 working days and lots of documents. The Chinese cruising passenger market in 2015 will account for 1+ mill – for the first time ever.

Visa-free entry in South Korea is norm since 2013 for all the major operators on the Asian market. The Korea and Japan visa-free cruise travel is a strong stimulant, since ~90 per cent of all China homeported cruise itineraries visiting Japan and South Korea will be visa free.

ASIA CRUISE MARKET TRENDS 2015

According to the Asian market's 2015 reports, major line companies have built up capacity in the region. The biggest player (for now) is Royal Caribbean. The RCCL company's Asia cruise market share growth is 50 per cent. Next are the budget lines Costa (35 per cent up in 2015) and Princess (15 per cent up in 2015).

The German line AIDA Cruises doubled its Asian market capacity in 2015, able to carry 20,000+ passengers there. The CLIA Asia's latest report shows that for the period 2012-2014, the Asian cruise industry has undergone an exponential 34 per cent growth in all aspects. The Asian trend features a substantial annual growth in ships number, itineraries and passengers. In 2014, Asia cruise lines hosted ~1,4 mill Asians. The global cruising industry now recognizes Asia as a major market. The Asia's passenger capacity between 2013-2015 increases by 20 per cent annually, with an expected 2015 capacity of almost 2,2 mill cruise passengers.

The number of Chinese cruise passengers in the period 2012-2014 grew 79 per cent per year, with a total of 697,000 pax from mainland China carried in 2014. This is almost the total capacity of the market, since all other markets in Asia combined service 701,000 pax. Statistics show that the largest Asian cruise markets are China, followed by Malaysia, Indonesia and the Philippines, with 4 out of 10 pax aged under 40. Asian passengers prefer to sail close to home, and 9 out of 10 passengers cruise within Asia. In 2015, the Asian market will be operated by 52 cruise line ships, offering 1065 cruises with scheduled visits to ports of calls in Japan (646), Malaysia (580), South Korea (377) and Singapore-Thailand (374 coombined).

ROYAL CARIBBEAN ASIA 2014 – SKYSEA CRUISE LINE

In September 2014, RCCL and Ctrip (Chinese travel agency) established a joint venture named "SkySea Cruises". SkySea starts operations in mid 2015, and targets the Chinese cruise market exclusively.

The new line begins with only one vessel – the former Celebrity Century ship, that was sold to Ctrip. Ctrip and Royal Caribbean each own 35 per cent of the SkySea. The balance of 30 per cent is owned by a private equity fund and the line's management.

CRUISE PASSENGER STATISTICS

By 2018, over 24 million cruise passengers will be carried worldwide. Almost 60 per cent of them will be North American, 27 per cent – European.

Cruise ship passengers annual number	20,335 mill
Carnival and Royal Caribbean passengers (Disney passengers)	~14 mill (0,5 mill)
Cruise passengers from North America Note: The US states producing the most passengers are: California, Florida, Texas, New York, Massachusetts, Pennsylvania, Illinois, New Jersey, Georgia and Arizona.	80%
Children 18- year of age cruising with their families	1,6 mill
Number of cruise passenger deaths since 1979 (all causes)	172
Passenger age: 1. 25-29 2. 30-39 3. 40-49 4. 50-59 5. 60+	1. 7% 2. 18% 3. 26% 4. 22% 5. 26%
Avrg cruise passenger age	47
Passenger race (black, white, other)	3-91-6%
Avrg annual household income 1. $39-50k 2. $50-60k 3. $60-75k 4. $75-100k 5. $100-200k 6. $200-300k 7. $300+k	$97000 1. 9% 2. 10% 3. 16% 4. 19% 5. 39% 6. 7% 7. 1%
College graduated passengers	71%
Married and full time working passengers (retired)	62% (20%)
Male-female (married)	51%-49% (80%)
Passengers age over 25 yo earning US$40,000 and over	44,6%
Cruise passenger percentage of total US population	19,9%
Avrg cruise passenger budget PP per week	$1770
Avrg budget PP per week on a land-based vacation	$1200
Avrg time between trips	~3,5 years
Passengers who think a ship cruise is the best way to sample destinations	80%
Passengers who would return to Caribbean on a land-based vacation	50%
Passenger cruising with the spouse	75%
Passenger cruising with children (18- yo)	25%
Passenger cruising with friends	23%
Passenger cruising with relatives	21%
Cruise passengers' top destination choices: 1. Caribbean 2. Alaska 3. Bahamas 4. Hawaiian Islands 5. Europe (Mediterranean, Black Sea) 6. Bermuda 7. Europe (Baltic, Norway Fjords, UK, Canaries, Iceland) 8. Panama Canal 9. Mexico Riviera	Percentage 1. 45 2. 24 3. 23 4. 15 5. 14 6. 11 7. 9 8. 8 9. 8.

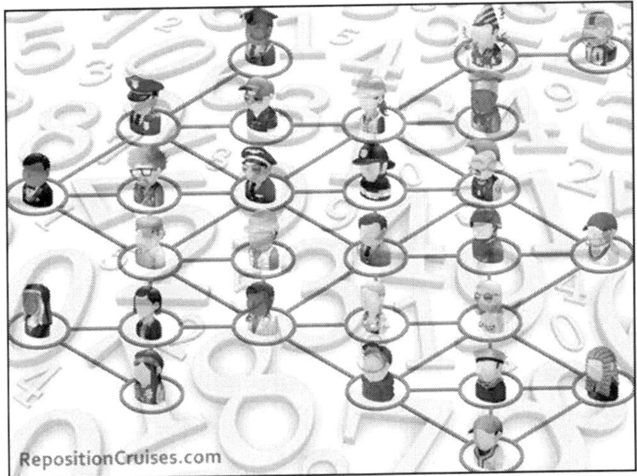

ROYAL CARIBBEAN ASIA 2016 SHIPS

In 2016, RCCL is repositioning in this region a fleet of 5 big-capacity vessels, including two new Quantum-class ships – Quantum of the Seas (2014) and Ovation of the Seas (2016). The five Royal Caribbean Asia ships in 2016 (Quantum, Ovation, Mariner, Legend and Voyager) will be homeported in 4 Chinese ports – Shanghai, Tianjin/Beijing, Xiamen and Hong Kong. They will operated 3- to 12-night long Asian cruise itineraries to ports in Japan, South Korea, Taiwan, Vietnam.

In addition to sending vessels to the region, the RCCL company is also investing in people on shore.

In August 2014, RCI created a curriculum with the Tianjin's Maritime College in order to attract, train and later employ Chinese onboard staff in the fields of culinary, restaurant and hotel services. As of early 2015, the RCI cruise line has hired over 3000 college graduates. RCI is also considering the usage of China's major drydock facilities for the refurbishment of the Legend ship in 2018.

CRUISE TRENDS 2015

Worldwide cruise industry trends involve lines building more and bigger ships, and, of course, cruise companies taking more of your vacation money. The list of most popular trends includes

- More and diverse dining options (alternative restaurants)
- Cheap tickets prices and not cheap alcohol/drinks
- Alluring package deals
- Designer goods shopping
- Faster Internet (shipwide Wi-Fi, mobile phone apps)
- New brand partnerships
- A higher demand for solo- and small-ship cruising deals.

CRUISE INDUSTRY TRENDS 2015

Why "industry trends" and not simply "cruise trends". Because the industry is all about the big money spend on ship travel vacations, and the trends are the big-time onboard spending to keep going – and faster.

More Specialty Restaurants with Surcharges

20 years ago, there were only main dining rooms as an onboard food option – sitdown dinner, no additional fees. Then the specialty restaurants came along, where a nominal cover charge PP applies. Followed the introduction of the fancy-dinner restaurants with a la carte pricing or per-person fee. The latest cruise trend – they now add surcharges on top of cover charges (supplements for some items). Among the lines that implement this strategy are Holland America (Pinnacle Grill), Royal Caribbean (Chops Grille), NCL Norwegian (Le Bistro).

Main Dining Rooms with Improved Experience

Main Dining Rooms offer the traditional fee-free fare on the ship. Carnival Cruise Lines introduced in 2014 the American Table family-style dining concept with a new menu. Portion sizes are larger, and table mates can share their food. NCL added the Tropicana Room on Norwegian Getaway – the dinner experience with live entertainment. Other lines (like Princess, Hurtigruten, Lindblad) upgraded the onboard cuisine choices with dishes depending on the ship's itinerary (destination's ports of call). This way, they source food (meat, fish, vegetables) from local producers and markets. Royal Caribbean introduced in 2015 the Dynamic Dining on the new RCI ships Anthem and Quantum. It means no traditional dining rooms, instead, passengers can dine when and with whom they want at several preset venues.

Alluringly Cheap Cruise Fares

Mainstream lines (CNL, MSC, Celebrity) begun to break down fares putting out some irresistibly low fares to entice customers on the ship, then start charging for optional "bundle" deals, providing a more personalized vacation experience. The list of such offers includes deals with airfare, hotel inclusive packages, all-inclusive drink packages (some with included Internet), etc.

Cruise Sales Booming

Cruise deals now look appealing the whole year-round. They are no longer reserved for peak season or holiday travel times, with booking windows shifting constantly. The major lines' reactions result in more promotional offers, with more diverse choice of bargain deals depending on value. Cruise passengers now can choose between promotions with bonus amenities or bargain cheap tickets prices.

Mobile Apps, Faster Cruise Ship Internet

In 2014, the cruise market's two biggest companies Carnival Corporation and Royal Caribbean started the battle for the "Internet at Sea" supremacy. RCCL/Royal Caribbean in 2013 partnered with O3b for an array of satellites providing faster Internet speeds on RCCL ships. Then Carnival Corp in 2014 implemented a hybrid system (land-based and satellite Wi-Fi systems) bringing the Internet speed up to 10 times faster compared to the industry's standard. These new systems are currently available only on the companies' newest ships, but the trend is to go fleetwide, and covering all the subsidiary brands.

In the begining of 2014, only few lines offered cruise mobile apps. By the 2014 end, all major companies offered at least such application, with most of them providing mainly pre-cruise research tools. Some apps feature also onship uses – from scheduled daily activities to making reservations (Spa, dinner) to connecting with other passengers. In 2015 this trend continues, with Carnival and Holland America launching their own mobile apps.

More Casual Alternative Restaurants

Alternative restaurants offer multi-course, gourmet meals, with traditionally over-the-top presentation, quality and taste. All that for a flat fee PP. Some of the major lines begun to create more casual specialty dining venues for superb food at a la carte pricing and more casual dress code. Some such venues are even charge-free. The list of the most popular ones includes Royal Caribbean's "Michael's Genuine Pub", the NCL Norwegian's pub "O'Sheehan's Neighborhood Bar and Grill", Princess Cruises' "Alfredo's Pizzeria" (free of charge). The trend is their number to grow, including new bars with gourmet quick bites and brew pubs with draft beers.

Cruise Shopping

Shopping at the onboard retail shops and boutiques goes high-end. This is probably great news for brand goods / expensive shopping afficionados. Now more and more ships have designer brands on sale with duty-free pricing. Celebrity added Cole Haan and Kate Spade to Michael Kors. Princess added Hugo Boss, Trina Turk, Diane von Furstenberg, Fendi, Salvatore Ferragamo. While this is done mostly for the Asian cruise market, the trend is these partnerships to be introduced on ships in other regions too.

Single Cruise Passengers Get More Attention

Solo cruising's top disadvantages are the higher cost (double occupancy rates) and the limited/small number of single cabins (aka Studio staterooms with single occupancy rates). The trend here is more solo cabins to be added – especially on the new vessels. NCL Norwegian, Royal Caribbean, P and O and Fred Olsen lead the way with a good number of single cabins on some of their ships. Holland America launches its newest ship ms Koningsdam (2016) with the line's first-ever single occupancy cabins. NCL expanded the Escape ship's Studio area from 59 rooms (on Breakaway class ships) to 82. The river cruise lines AmaWaterways and Tauck offer solo travel cabins on some of their riverboats. And many of the new river ships are now equipped with single staterooms.

Small Ship Cruises on the Rise

Windstar Cruises doubled their fleet by buying three of the Seabourn yachts. Compagnie du Ponant lines introduce their newest ship (264-pax capacity) in 2015. Oceania bought the Ocean Princess ship and renamed it ms Sirena (scheduled debut in 2016). Viking Cruises launched the new Viking Ocean brand to offer sea cruise of three identical-design cruise vessel. The trend is, with so many huge ships, the demand for and popularity of small ship cruising vacations to continue to grow.

The cruise market is booming, with more and more opportunities for the largest companies to make huge profits depleting your sweet vacation money fund. Nowadays, ship cruising looks more and more like land-based resort vacations, and the number of small ships in operations continues to decline. Still, cruise travel is the cheapest, most intriguing and most-fun way to spend time and money on seasonal vacation, short-break or during holidays. Enjoy it!

ECONOMIC GROWTH OPPORTUNITIES

Fig. Cruise Ships at Port of Saint John, New Brunswick.

International trade fuels economic growth. Each $1 billion in exports generates as many as 11,000 new jobs. The Atlantic Gateway and Trade Corridor strategy will build on the region's existing businesses and technologies, its natural assets and resources, its skilled labour, its reliable transportation infrastructure and its wide market reach to facilitate international trade and value-added services.

Cruise Line Partnerships

Disney Cruises started this game, but themed children's clubs are not a novelty at sea. However, more and more lines introduce the themed character experience for kids, with more and more such partnerships being announced. In 2015, Carnival added fleet-wide their "Seuss at Sea" programme. MSC teamed up with LEGO introducing LEGO-playrooms on the Armonia ship and also on the line's two newbuilds (debuting in 2017). MSC also partnered with Chicco providing baby gear to families. Royal Caribbean partnered with Dreamworks, NCL Norwegian with Nickelodeon.

INTERNATIONAL TRADE

The Atlantic region currently handles significant levels of international trade, with some $43 billion worth of goods moving through the system in 2009. Of this, exports contribute to approximately one third of the region's GDP. Regional businesses have strong export markets in refrigerated (reefer) cargo, agricultural, forestry and energy products, minerals, and in defined manufacturing sectors. The Atlantic Gateway and Trade Corridor Strategy will help these exporters increase their global reach and competitiveness and generate back haul opportunities for international shippers using Canada's east coast ports. The Atlantic Gateway and Trade Corridor strategy will also enhance export capacity in the region by helping export- ready businesses enter international markets. Trade promotion activities will leverage existing export activity by opening new markets and strengthening links with key existing markets such as the United States.

Planned activities will:

- Support integrated trade missions;
- Develop and implement sector export strategies; and
- Provide export counseling and mentoring programmes and services to enhance export preparedness.

Significant quantities of energy products move to and from the Atlantic region. Over 90 percent of Atlantic Canada's energy exports go to the United States. In 2008, these exports totalled $19.2 billion.

In the future, the demand for transporting crude and refined petroleum products is expected to grow because of:

- Increased domestic production off the coast of Newfoundland and Labrador;
- Increases in refining capacity in Saint John and Placentia Bay; and
- Liquefied natural gas operations.

There is also a current thrust to grow the energy sector by focusing on generating and exporting electricity from renewable sources, including hydro, wind, tidal, nuclear and biofuels. The Atlantic Gateway and Trade Corridor Strategy will help further develop an energy sector in the region through stronger transportation and logistics systems, improved export capacity and a skilled, knowledgeable and innovative workforce.

VALUE-ADDED SERVICES

Nearly 44,500 Atlantic Canadians are directly or indirectly employed by transportation and logistics businesses, generating $6.2 billion in economic output (Atlantic Gateway Business Case, InterVistas, 2007). Many of these businesses add value to products along the Atlantic transportation corridor by handling, processing, screening, sorting, warehousing and transporting cargo. Private businesses and regional port and airport authorities have state-of-the-art transload, distribution and warehousing centres. These value-added services connect the region's transportation network to global supply chains, ensuring that business is done in the region rather than simply through the region.

Within the Atlantic Gateway and Trade Corridor Strategy, governments and the private sector will identify sectorbased opportunities to provide value-added services within the region. For example, given its proximity to existing shipping lanes and air access capabilities, Newfoundland and Labrador businesses may be positioned to support transload, offshore supply and service, and other commercial activities in the North.

CRUISE

In Atlantic Canada, the cruise industry is a major contributor to the region's economy, particularly the tourism sector, with an estimated total economic impact of $175 million in 2008. Overall, the region has seen strong growth as its share of the North American cruise market has risen from 1.54 percent in 2005 to 2.11 percent in 2008. Passenger throughput has risen 128 percent over the last ten years. Looking to the future, Atlantic Canada stands to benefit from positive forecasts for the international cruise market (for example, growth in passenger levels and vessel size) generating new homeporting and port-ofcall opportunities. Cruise passenger visits typically generate further economic opportunities for the tourism industry through extended stays and follow-up visits to the region.

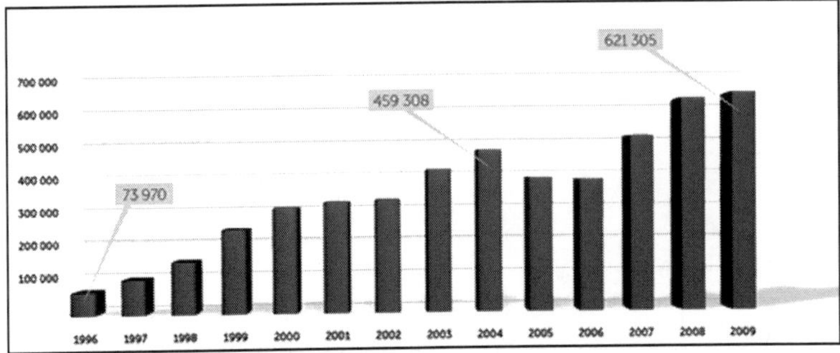

Fig. Growth in Cruise Passengers to Atlantic Canada.

Planning within the cruise industry of specific itineraries takes place well in advance of implementation with consideration given to how a system of ports in a specific region can offer an attractive destination for potential cruise passengers while maximizing economic returns. The availability of shore side infrastructure, comprehensive shore excursion packages and support for other land based tourist activities play a large role in influencing planning decisions within the industry. Consequently, competing effectively for cruise business requires an ongoing collective approach from key players in the region.

IMMEDIATE MEASURES

Halifax Stanfield International Airport Runway Extension (NS) - to extend runway to accommodate larger cargo planes and enhance safe operations. Greater Moncton International Airport Runway Extension (NB) - to extend runway to accommodate larger cargo planes and enhance safe operations. Port of Saint John Cruise Expansion (NB) - to expand cruise berth and passenger handling capacity. Charlottetown Airport Terminal Expansion (PEI) - to capitalize on economic growth and facilitate increased traffic.

ASIAN CRUISE MARKET GROWS AS CRUISE LINES EXPAND OFFERINGS

Fig. This is Part One in a Two-Part Series on Asian Cruising.

It wasn't so long ago that the Asian cruise market was perceived by many as inconsequential compared to the Mediterranean and Caribbean markets. But according to a recent survey from CLIA, Asia is quickly becoming a major player in the cruise industry as lines shift capacity to meet a growing demand from Asian cruisers. Two years after Carnival Asia CEO Pier Luigi Foschi predicted Asia would account for 20 per cent of the world cruise market by 2020, the region has increased its capacity share almost double any other region, and now owns the fourth largest cruise capacity—behind the Mediterranean, Caribbean and Europe—of the regions CLIA measured. "The cruise vacation market in China is still small, relatively speaking. However, it is the fastest growing market in the world," said Dwain Wall, co-president and COO of

WorldCruise.cn, a China based cruise-booking and information web site. Wall, who is a former CLIA senior vice president, said that there is tremendous "untapped potential" in China, both for agents and travelers.

INDUSTRY REACTION

As the number of people taking cruises in Asia has increased, so has the involvement from major cruise lines. From 2013 to 2015, the number of ships in the region increased from 43 to 52 as more cruise lines moved ships to Asia. Most notably, Royal Caribbean's Quantum of the Seas relocated from Bayonne, N.J., to Shanghai in 2015, something Adam Goldstein, president and COO of Royal Caribbean Cruises, said was an effort to "accelerate the growth of this vital market." Princess Cruises has also increased their presence in the region, including earmarking their new ship set to debut in 2017 for Asia. "Deploying our next new ship in China underscores our strong commitment to growing the China cruise market and providing discerning travelers with amazing vacation experiences at sea," said Jan Swartz, president of Princess Cruises.

THE NEW ASIAN CRUISER

According to CLIA, most of the increase in interest for Asian cruising has come from inside the region. In 2014, cruise lines hosted 1.4 million Asian guests, up from 774,000 guests in 2012. And while more passengers from all over Asia are choosing to cruise, the most significant population has come from China. The population of Chinese cruisers is almost ten-times larger than any other Asian country. While cruising is still a relatively new concept to the population, cruises have become attractive to Chinese customers because of an easing of visa restrictions and a burgeoning middle class with more disposable income. The Chinese cruiser is both younger and seeking shorter cruises, according to Wall, than the North American cruiser. "This [under 40] age group is much more likely to try cruising," said Wall. "The three- and four-day market is popular, as Chinese can experience cruising at a lower cost and without having to be away from home or work for an extended period of time."

EFFECT ON AGENTS

Even though the market is driven by the Asian customer, a higher commitment from cruise lines to the region means more opportunities for sales. Because of the region's size and differences in culture and attractions, Asia is region made for cruising, especially for clients who are seeking destinations outside of the Caribbean and Europe. "I would suggest that agents promote a cruise vacation in China and Asia as a way for travelers to experience multiple destinations without the hassles and expense of flying from city to city...Emphasize the affordability, the convenience and the exotic nature of the sights and cultures they will experience," Wall said.

Bibliography

Geetanjali.: *Tourism Geography*, Centrum Press, Delhi, 2010.

Manish Ratti.: *History and Geography of Tourism*, Rajat Publication, Delhi, 2007.

Rana Pratap and Kamla.: *Prasad Tourism Geography*, Shree Publication, Delhi, 2005.

Subhash Chandra Sharma.: *Geography of Tourism (2 Vols-Set)*, Rajat Publication, Delhi, 2002.

A.K. Sarkar.: *Action Plan and Priorities in Tourism Development*, Kanishka Publication, Delhi, 2010.

Alan A. Lew, C. Michael Hall, and Allan M. Williams.: *A Companion to Tourism*, Rawat Publication, New Delhi, 2005.

Anupama Srivastava and Keya Pandey.: *Anthropology and Tourism*, Serials Publications, Delhi, 2012.

Anurag Kothari.: *A Textbook of Tourism Management*, Wisdom Press, Delhi, 2011.

Arvind Gautam.: *Critical Analysis of Hospitality and Tourism Industry*, Axis Publication, New Delhi, 2010.

B. K. Goswami and G. Raveendran.: *A Textbook of Tourism*, Haranand Publications, Delhi, 2010.

Babu P. George and Sampad Kumar Swain.: *Advancements in Tourism Theory and Practice : Perspectives from India*, Abhijeet Publication, Delhi, 2005.

Cynthia vanden Driesen and Satendra Nandan.: *Austral-Asian* Encounters : From Literature and Women's Studies to Politics *and Tourism*, Prestige Books, New Delhi, 2003.

David Carr.: *Community Tourism and Natural Resource Conservation*, Discovery Publication, Delhi, 2011.

Debasish Mazumdar and Lavkush Mishra.: *Contemporary Tourism Development: Issues and Challenges*, Rajat Publication, Delhi, 2010.

Dileep Makan.: *Conceptualization of Tourism*, Adhyayan Publication, New Delhi, 2006.

Dilip Das.: *Critical Issues in Tourism*, Murari Lal and Sons, Delhi, 2011.

Graham Dodgshun and Michel Peters.: *Cookery for the Hospitality Industry*, Cambridge University Press, New York, 2001.

Hemant Sharma.: *HRM in Hospitality Industry*, ABD Publication, Delhi, 2006.

Jack Randall.: *Agriculture Tourism*, Discovery Publishing House, Delhi, 2011.

Jitendra K. Sharma.: *Contemporary Tourism and Hospitality Management*, Kanishka Publication, Delhi, 2006.

N. Jayapalan.: *An Introduction to Tourism*, Atlantic Publication, New Delhi, 2001.

P.K. Bal.: *A Text Book of Hospitality Tourism and Aviation*, Cyber Tech Publication, Delhi, 2011.

Percy K Singh.: *HRM in Hotel and Tourism Industry : Existing Trends and Practices*, Kanishka Publication, Delhi, 2008.

Pragati Mohanty.: *Hotel Industry and Tourism in India*, APH Publication, Delhi, 2008.

Prateek A. Aggarwal.: *Aspects of Crosscultural Interaction and Tourism*, Mohit Publication, Delhi, 2005.

Prem Nath Dhar.: *Cultural and Heritage Tourism : An Overview*, Kanishka Publication, Delhi, 2008.

Ramesh Raj Kunwar.: *Anthropology of Tourism : A Case Study of Chitwan-Sauraha, Nepal*, Adroit Publication, New Delhi, 2002.

Rattandeep Singh.: *Commonwealth Games and Sports Tourism : Global and National Perspectives*, Kanishka Publishers, Delhi, 2010.

Ravee Chauhan.: *Advanced Book on Marketing of Tourism*, Vista International Publishing House, Delhi, 2011.

Romila Chawla.: *Accommodation Management and Tourism*, Sonali Publication, Delhi, 2006.

Romila Chawla.: *Agri-Tourism*, Sonali Publication, Delhi, 2006.

Romila Chawla.: *Coastal Tourism and Development*, Sonali Publication, Delhi, 2004.

Rowe.: *Career Award Travel and Tourism: Standard Level*, Cambridge University Press, New York, 2003.

S W P Prabhakaran.: *Child Labour In Hotel Industry*, Discovery Publishing House, Delhi, 2011.

Saurab Kumar Dixit.: *Aspects of Tourism Development*, Mohit Publication, Delhi, 2005.

Suddhendu Narayan Misra and Sapan Kumar Sadual.: *Basics of Tourism Management*, Excel Books, New Delhi, 2001.

Index

I

L

M

N

O

P

Q

R

S

T